CRISIS COMMUNICATIONS: A PRIMER FOR TEAMS

CRISIS COMMUNICATIONS: A PRIMER FOR TEAMS

ROLES • RESOURCES • PROCESSES • PRINCIPLES

BY AL CZARNECKI, APR

iUniverse, Inc.
New York Lincoln Shanghai

Crisis Communications: A Primer for Teams
Roles Resources Processes Principles

Copyright © 2007 by Al Czarnecki, APR

All rights reserved. No part of this book may be used or reproduced by any means, graphic, electronic, or mechanical, including photocopying, recording, taping or by any information storage retrieval system without the written permission of the publisher except in the case of brief quotations embodied in critical articles and reviews.

iUniverse books may be ordered through booksellers or by contacting:

iUniverse
2021 Pine Lake Road, Suite 100
Lincoln, NE 68512
www.iuniverse.com
1-800-Authors (1-800-288-4677)

This book represents my best advice on crisis communications. While much effort went into making it accurate, well considered, and complete, advice is merely a recommendation—something presented as worthy of acceptance. It should not be followed blindly or without further consideration. In the end, you must be prudent and make your own decisions. I wrote this book to assist teams with that process.

Al Czarnecki, APR
Toronto, Canada

ISBN-13: 978-0-595-40613-5 (pbk)
ISBN-13: 978-0-595-87750-8 (cloth)
ISBN-13: 978-0-595-84980-2 (ebk)
ISBN-10: 0-595-40613-0 (pbk)
ISBN-10: 0-595-87750-8 (cloth)
ISBN-10: 0-595-84980-6 (ebk)

Printed in the United States of America

Contents

ILLUSTRATIONS.................................... vii

INTRODUCTION..................................... ix

Part 1 THE TEAM

CHAPTER 1   KEY PLAYERS AND THEIR ROLES 3

Part 2 THE CRISIS SOUP

CHAPTER 2   SCENARIOS 15

CHAPTER 3   RESOURCES 25

CHAPTER 4   ROLES 43

CHAPTER 5   PROCESS 52

CHAPTER 6   PRINCIPLES 63

Part 3 ISSUES AND ACTION

CHAPTER 7   ISSUES MANAGEMENT 69

CHAPTER 8   EMOTIONAL INTELLIGENCE 79

CHAPTER 9   FOCUS 84

CHAPTER 10   NEWS CONFERENCES 92

CHAPTER 11   MEDIA ACCURACY 107

CHAPTER 12   TESTING 115

Part 4 RESILIENCE AND CONTINUITY
CHAPTER 13   EMERGENCY PROVISIONS 123

Part 5 DEVELOPMENT
CHAPTER 14   BUILDING YOUR TEAM. 149

APPENDIX A   SUPPLEMENTARY MATERIAL. 155
APPENDIX B   LINKS TO WEB-BASED RESOURCES 171

ILLUSTRATIONS

1. Key players 4
2. The crisis soup (mind map)* 16
3. Readiness checklist 28
4. Roles 45
5. Process 58
6. Principles 65
7. Crisis Dos and Don'ts (wallet card)* 114
8. Resilience 144
9. Media request to speak with families (form)* 156
10. Story Action Sheet (form)* 157
11. Microcassette counter vs. minutes (chart)* 158

*Available as a printable PDF at http://topstory.ca/crisisteambook.html. To enter the private area you'll need a username and password. You'll find these on the last page of this book.

INTRODUCTION

Less than a decade ago, crisis communications usually meant you faced a threat specific to your organization. Today, crises and disasters related to severe weather or biological or political threats are top of mind. There is a wider mandate for both response and communication.

This book has been written for three main audiences:

- Senior managers who want a new tool to develop their crisis response team
- Organizations without an accredited public relations professional
- Communicators looking for new ideas on crisis communications

This book is intended to be a succinct, interesting, and useful read for busy people.

Part 1, "The Team," outlines the salient roles of key players prior to and during a crisis situation. Use this concise chapter and the table of contents to engage your team in reading and discussing this book.

Part 2, "The Crisis Soup," describes five aspects of planning for crisis communications: scenarios, resources, roles, process, and principles. The last four items strengthen your organization's resilience.

- *Scenarios* drive your crisis communications planning. Everything swims in the scenario; it is the water of the crisis soup.
- *Resources* are what you have on hand. They can be prepared in advance. If you can pull something useful out of a bag, you save time better spent on more urgent things.

- *Roles* define who does what and when. Crises involve multiple teams working together—your own and those of other organizations. Knowing who will field which issues helps everyone focus on his or her proper work.

- *Process* is the critical path for responding to a situation. It has a sequence and timelines. There are first steps, next steps, and activities done in parallel. In the blur of a crisis or disaster, process helps teams work effectively.

- *Principles* are standards that apply to all your actions. When everything seems to be up in the air, clear values help you quickly decide what's right.

Part 3, "Issues and Action," covers operational details relevant to crisis communications.

- *Emotional Intelligence* offers seven quick takes on feelings as important facts.

- *Issues Management* deals with situation monitoring and managed interdependence.

- *Focus* addresses priorities, positioning, and messages.

- *News Conferences* covers news and the process of releasing it, both externally and internally.

- *Dos and Don'ts* is a short list of nine important things to do and nine things to definitely not do.

- *Testing* cites six reasons to test your plan and offers some suggestions.

Part 4, "Resilience and Continuity," considers how to keep your team functioning throughout a disaster. *Emergency Provisions* outlines down-to-earth preparations for even the smallest organization.

Part 5, "Development," suggests how to move forward on crisis communications readiness. *Building Your Team* outlines a process for developing senior manager involvement.

Appendix A offers some sample documents. Appendix B contains selected and annotated URLs. You can download a bookmarked PDF file of Appendix B with live links to all the URLs. The username and password to access the download area are on the last page of this book.

Terms

It's helpful to define some terms at the start—crisis, disaster, catastrophe, public relations, stakeholders, and publics.

A *crisis* is a time of intense difficulty or danger, a compelling situation that demands immediate response. There are two basic types of crises—organizational and community:

- Organizational crises usually involve the motives of money, power, or sex.
- Community crises usually involve severe danger that has only recently become apparent.

Both organizational and community crises require urgent responses.

A *disaster* is an accident or a natural event that causes great damage or loss of life, affecting an entire community. Every country has its own list of disasters within recent memory. Searching for *disaster* on the Internet will bring up hundreds of examples, spanning centuries, from all areas of the globe.

A *catastrophe* is a disaster that overwhelms a community's ability to respond. Two examples, almost a century apart, are the Halifax Explosion of 1917 and the Indian Ocean Tsunami of 2005.

The communications function in many organizations focuses primarily on marketing.

The agendas of *public relations* and marketing are different. Marketing is interested in the market—consumers and demand. Public relations is interested in relationships—reducing conflict and improving cooperation. Marketing adds value by increasing income. Public relations adds value by decreasing expenses that become necessary when issues have been ignored.

The words *stakeholders* and *publics* are often used interchangeably, but they have more precise meanings.

Stakeholders are groups with an interest or investment in an issue. *Publics* are audiences defined by their response to an issue. The same people, but grouped differently, are often involved.

Publics form themselves as people become concerned about, develop feelings regarding, and take action on something that matters to them. This naturally involves stakeholders. Public relations fosters two-way communication to help organizations and their publics adapt to one another, forming relationships.

Wicked situations

Crises often contain elements of what management science calls a "wicked" problem. Wicked problems are characterized by:

- Inadequate information
- Conflicting objectives
- More than one decision maker
- A turbulent environment
- Several interwoven problems
- Costly and irreversible solutions

Good planning, coordination, and communication can mitigate the first three elements. The final three are inherent in many crises and are sufficient challenges in and of themselves.

Information gathering, issues management, and due diligence are all elements of crisis communications.

A crisis can sometimes be prevented from progressing to a disaster or a catastrophe. If a crisis cannot be prevented, its impact can often be reduced.

Case studies

This book focuses on principles, process, and provisions. Its purpose is to facilitate discussion and act as a resource for management teams as they develop their crisis communications plan.

My priority has been to create a lean, fast read for busy people. There are URLs in Chapter 2 that point to a range of crisis scenarios. At the end of Appendix A is a short list of books that contain case studies.

URLs in PDF file for download

Some highly useful resources are available on the Internet. I've simply listed them in this book, then packaged them in a PDF file that's available for download. (See Appendix B for URLs)

The PDF with URLs is located at http://topstory.ca/crisisteambook.html. To enter the private area you'll need a username and a password. These are shown on the last page of this book.

You'll need Adobe Acrobat Reader to view this PDF file. This software is freeware and comes installed on most computers. If you need to download a more current version for either Macintosh or Windows systems, go to http://www.adobe.com/products/reader/.

The PDF of Appendix B has bookmarks in the left column. Beside each chapter is a triangle icon. When you click on this, it pivots and subtopics drop down. URLs in the book are listed by subtopic. All links are "live" (underlined and blue). Keep the PDF open while reading the book, and you can just click through to visit cited URLs.

I will post an updated PDF file at my Web site each year during the first week of January. I can do this only for a limited time. So if you're reading this book five years from now, here are some tips.

There are a number of reasons why a link may not work:

- A busy server may be down for maintenance, or your connection may have timed out.

- The base URL may have changed, or the site may have reorganized its folders.

- The site or the page is no longer available.

It can be worthwhile to try again a few hours later or the next day. If that doesn't work, try a Boolean search that includes the base URL (the part between http://and the next slash) in quotes AND a keyword or phrase. You want to include words that are certain to be on the Web page, while limiting results.

How to read this book

Read through the chapters quickly, discussing them with fellow team members. You may want to organize breakout sessions so that different groups can discuss topics specific to their roles.

Go back to the chapter on scenarios and choose one or two for a planning exercise. Use the book and the accompanying PDF file as resources in putting together your own crisis communications plan.

Your team can read and discuss this book in twenty hours. A typical work year has 250 days, so this represents about a 1 percent investment in time.

Formal component and time required:

- You can read each chapter in twenty to thirty minutes (just four to six hours for the entire book).
- Follow up the reading with forty to sixty minutes of team discussion on each chapter (ten to fourteen hours in total).

Informal component:

- Browse Web resources listed in the accompanying PDF file.
- Discuss a topic in detail over lunch.
- Incorporate agenda items into regular management meetings.

Each organization has its own corporate culture and working style. A weekly meeting could cover the entire book in three months. Alternatively, coverage could be compressed into several half-day sessions.

Follow-through on discussion and decisions merges into existing priorities and regular management time.

Feedback

I welcome hearing from you about this book—what you have been able to put to use and your suggestions for making the next edition better suited to your needs.

My e-mail address is ac@topstory.ca. For safe passage through my SPAM filter, include the word FEEDBACK in the subject line.

I can't promise that I will respond to every e-mail, but I will read and carefully consider each one. I hope you find this book both interesting and useful.

—Al Czarnecki, APR

PART 1
THE TEAM

Trust and group dynamics are important in a crisis situation. People have only the resources at hand, their wits, and each other. It's important for teams to have asked the right questions and discussed their answers—before they are face to face with a crisis. This book is designed to be a catalyst for that process.

1
KEY PLAYERS AND THEIR ROLES

In a film on shipwrecks, the captain of a supership says that, despite sophisticated electronic systems, when you're at sea, you're on your own—and if you want to survive, you'd better have a team.

The advice also holds for crisis situations.

Complacency based on presuppositions will be your first challenge in preparing for a crisis:

- Something like this will never happen.
- If it does, it won't involve my department.
- Our response can be put together in real time.
- The media will wait.
- Our reputation, our brand, our relationships won't be affected.

Asking the right questions is the shortest path to the right answers. This book will help your team examine and discuss their respective roles in a crisis or disaster. It will help them prepare useful resources and understand how each department supports others.

◆   ◆   ◆

Here's an outline of the typical players in a crisis management team and their salient roles prior to and during an incident. When a crisis occurs, they look to one another and pull together as a team.

CEO • BC • HR • IT • PR • FM • Department Heads

CEO—chief executive officer

- Sets priorities
- Allocates resources
- Brings everyone to the table
- Ensures accountability, due diligence, transparency
- Ensures that all employees take crisis preparation seriously
- Brings all resources to bear on dealing with the crisis

BC—business continuity manager

- Works with all units to understand interdependencies and the impact of interruptions
- Ensures that contingency plans are in place in all units of the organization
- Ensures that plans have been tested, communicated, and are understood by senior management
- Ensures that a coordinated emergency response plan is in place
- Coordinates emergency response activities across departments
- Carries out regular updates and ensures that necessary training is done

HR—human resources manager

- Ensures that new employee orientation covers the crisis/disaster response program
- Cross trains for resilience and flexibility; sources replacement staff within a crisis scenario
- Develops policies, procedures, and basic training for disease prevention
- Develops policies for exceptional sick leave to support self-screening during a pandemic
- Develops work-at-home capability, policies, procedures
- Deals with refusals to work during a crisis or disaster situation

IT—information technology manager

- Ensures that all systems are robust and secure, and performs a periodic security audit
- Manages backup strategy, including off-site copies and periodic testing to restore

- Develops strategies for dealing with an extended power outage
- Develops alternatives in the event that some ISPs are out of service
- Develops capability for urgent Web site changes to support communication during a crisis
- Plans for alternate support in the event that regular IT staff are unavailable

PR—public relations manager

- Explains the role of public relations to new employees during orientation
- Fosters two-way communication to mitigate issues before they become crises
- Develops trust and credibility with employees, key stakeholders, and the media
- Promotes a common understanding of the organization: its mission, operations, role
- Ensures that all components of crisis communications readiness are in place
- Leads and coordinates the communications function during a crisis

FM—facilities manager *(at crisis location)*

- If appropriate, makes facilities available in response to a community crisis
- Deals face to face with neighbors, local leaders, and responders
- Acts as liaison between emergency personnel and the crisis management team
- Ensures location security at appropriate level

- Implements screening and quarantine protocols as appropriate
- Otherwise, role is similar to that of a department head (see next)

DH—department heads *(full senior management team)*

- Implement preparedness with their own families to ensure their own availability and focus during a crisis
- Ensure their own departments are fully compliant with crisis/disaster plans
- Streamline communication between the incident response team and frontline staff
- Actively engage employees in two-way communication to answer their questions
- Actively support responding departments, finding ways to ease their load
- Acknowledge and thank everyone who is helping with crisis response

The senior management team

In "Critical Infrastructures under Threat: Learning from the Anthrax Scare" *(Journal of Contingencies and Crisis Management,* September 2003), the authors (Boin, Lagadec, Michel-Kerjan, and Overdijk) raise the issue of critical infrastructures—the networks that facilitate traffic, financial transactions, communication, and the delivery of water, electricity, gas, and food—and the crises that threaten to disrupt them. A small glitch in one network can quickly cascade into large-scale breakdowns in other networks.

Among the examples offered:

- 1995 Kobe earthquake destroyed a harbor, contributing to the Asian monetary crisis of 1997

- 1998 Canadian Ice Storm deprived three million people of electricity for weeks, causing shortages in fuel, water, heat, and supplies

- 2001 Anthrax scare affected the postal networks of both North America and Europe

The impetus for this article was the 2002 Paris Conference where top-level postal executives gathered in the wake of the Anthrax scare. They launched a learning process to share experiences and lessons and increase their collective capacity to respond to similar events. (See Appendix B for URLs)

> "It is crucial ... that the administrative elites of public and private companies begin to understand that crises tend to be rapidly emerging and evolving processes that can turn into vicious and unmanageable circles."

The authors identify four patterns in current crisis management practices that are cause for concern:

1. Crisis managers tend to focus on hardware and ignore the "soft" human element. This focus on prevention can undersell the importance of resilience in responding to a crisis.
2. As challenges become systemic, local authorities are forced into unknown theatres with different actors, where they can find themselves feeling powerless.
3. A long-standing reliance on rational planning can be a liability. In systemic disruptions, the basic references experts use are often shattered. The situation requires urgent troubleshooting and new solutions, often outside the box.
4. Crisis management planning is too often just a paper exercise, leading to the false belief that the organization is well prepared. Plans must be tested to determine whether they will hold up in the actual event.

The article identifies three requirements for responding to the new dynamics of crises:

1. Crises cannot be simply delegated to technical teams; they require the involvement of the senior management team.
2. Crisis management must be based on the concept of resilience: learning to organize for the unknown.
3. Managers need to look to other organizations and networks and learn from their experience.

The final recommendations might represent a cultural revolution in many organizations:

- Involve new stakeholders, develop relationships, learn about other organizational cultures.
- Adapt your own communication culture to this learning, and nurture collective processes.

- Develop teams that advise top leaders, suggest innovations, engage with outside organizations.

- Organize structural debriefings for directors that surpass mere technical feedback.

- Run simulation exercises, followed by rigorous debriefings.

- Introduce training programs to develop both a generic crisis culture as well as expertise specific to the crisis function of individual leaders—three critical roles: leaders, crisis team facilitators, and strategic observers who monitor the crisis and report to the strategic level.

Everything the article recommends points to the function of senior management teams. The full text of this 4,600-word article from the *Journal of Contingencies and Crisis Management*, with a list of references, is available online. (See Appendix B for URLs.)

It's likely you'll need to deal with several organizational crises before you ever find yourself in a disaster situation. Developing a philosophy, competencies, systems, and resources will make your organization (and by extension your community) more resilient. This is an investment rather than an expense.

Problem-based learning

Problem-based learning (PBL) is a process in which:

- People assess what they know and what they need to know, then gather information and collaborate to solve a problem.

- It begins with an ill-structured problem that requires probing issues and dealing with complexity.

- The definition of the problem can change midstream (common in a turbulent environment).

- You must make important decisions based on the information at hand.

All these points are aspects of a crisis or disaster situation. PBL is oriented to the real world, requires team effort, and builds on existing knowledge and strengths. It can parallel and help develop a natural workplace process.

Samford University has a Web site on problem-based learning that includes links to resources.

(See Appendix B for URLs)

PBL requires a leader or facilitator and resources. You choose your leader. This book can serve as a key resource. The PDF file that accompanies this book offers a wealth of additional Web-based resources.

◆   ◆   ◆

Key Players and Their Roles—Questions for discussion:

1. Is the full senior management team clear on everyone's specific role during a crisis?

2. In a crisis, where might we find ourselves and with which other actors?

3. Would it be prudent to build some new strategic relationships?

4. Would it make sense for all our key players to read and discuss this book?

5. If so, how can we lead from the top and involve senior management?

6. Will the CEO participate and fully support this process?

7. How will we organize reading and discussing this book?

◆　　◆　　◆

ENDNOTE

The senior management team section of this chapter is adapted from Boin, A.; Lagadec, P.; Michel-Kerjan, E.; Overdijk, W. "Critical Infrastructures under Threat: Learning from the Anthrax Scare." *Journal of Contingencies and Crisis Management*, vol 1, no 3, September 2003: 99-104(6). Blackwell Publishing, 2003 with permission of the authors.

PART 2
THE CRISIS SOUP

A crisis is a turbulent environment. Important decisions with far-reaching consequences must be made quickly. Information is inadequate. The situation is constantly changing. The picture is much like a bubbling pot of soup: ingredients dodge and swirl, rise as priorities, and then recede as they swim in the issue's broth.

2
SCENARIOS

On September 13, 2006, twenty students were shot when a gunman went on a rampage at Dawson College in Montreal. A CBC Radio reporter was on the scene within ten minutes. Wire services picked up the story. Twelve hours later the incident held the top spot on Google News, showing 869 stories. Within twenty-four hours, coverage peaked at more than 1,200 stories worldwide.

Among the hundreds of media outlets that carried the story were Aljazeera.net (Qatar), BBC News (UK), *Daily News & Analysis* (India), *Gulf Daily News* (Bahrain), *International Herald Tribune* (France), *Kazinform* (Kazakhstan), *New York Times* (US), *News24* (South Africa), *People's Daily Online* (China), *Swissinfo* (Switzerland), *Sydney Morning Herald* (Australia), *The Globe and Mail* (Canada).

Five days later, Canadian media carried accounts of how the college was returning to normal classes.

Random and tragic violence is just one—thankfully rare—type of crisis. Whatever your crisis, there will be demands from the situation itself, the stakeholders involved (including families), and the media.

History's lessons

Just as necessity is the mother of invention, scenarios drive your planning process. Your crisis or disaster situation will be unique. Putting yourself in a past situation gives you clues to help you develop the resources, competencies, and teamwork to address what comes your way.

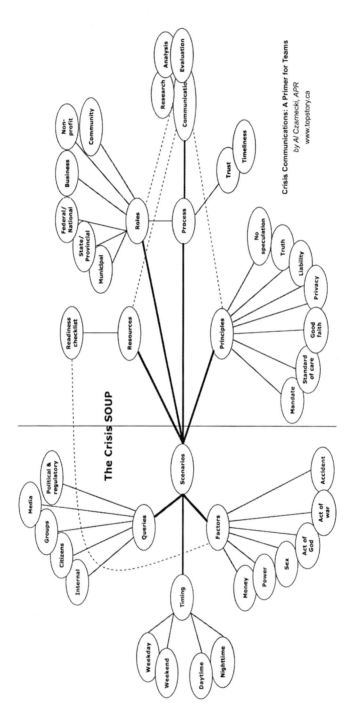

It's important to remember that any scenario is just a question. It can generate a variety of responses. There is often more than one correct answer. Your team needs to find its own.

Resilience will become a theme.

Getting started

The first priority in communications planning is to overcome inertia, to get things rolling, and to build some momentum within your organization. Start simple. Test everything. Make it work. Celebrate some success.

Acknowledge the good work people have done. Then raise the bar a little, incorporating what you've learned. Add a new, different scenario. Continue to challenge your team. Create scenarios that are complex and three-dimensional. Include unexpected complications.

Crisis response can work somewhat like method acting. It's not a matter of reading from a fixed script. Who you are determines how you act. This is where a values-driven approach empowers people. You let your characters find their lines as they see the play unfolding. Flexibility is important. They can depart from the script if doing so both supports core values and is more effective.

Some major types of crisis events:

- Health (Legionnaires' disease, SARS, an influenza pandemic)
- Major accident (Exxon Valdez, Mississauga derailment)
- Natural disaster (Hurricane Katrina, Great Ice Storm of 1998)
- Terrorism (9/11, Madrid and London subway attacks)
- Energy (East Coast power outage of August 2003)

- IT (a restore from backup destroys your system)
- You can add one top-of-mind risk that's especially relevant for you

Some online resources to help in developing scenarios

(See Appendix B for URLs)

Disasters—broadcast archives

- CBC Archives—Disasters
- CBS News—Disaster links
- BBC—Natural disasters

Disaster information

- FEMA Disaster fact sheets and backgrounders
- CDD—Emergency preparedness and response
- Worst United States disasters
- Canadian disaster database—information on over 700 events
- Natural hazards of Canada—150-year perspective
- Societal aspects of weather
- Engineering disasters—selected Web sites

(See Appendix B for URLs)

Organizational crises

Even ethical and accountable organizations can find themselves in a crisis:

- Your IT system goes down and five backups all fail to restore.
- A trusted employee is charged with embezzling a six-figure sum.

- The estranged husband of an employee walks in and kills her at work.
- A highly contagious disease causes your main location to be quarantined.
- You're doing everything right and a serious incident still happens.

Some crises involve a breach of standards or trust. People do, unfortunately, succumb to money or power or sex. Others in the same industry or profession may be implicated. Public discourse on the issue can involve these parties and government and advocacy groups.

Some online resources:

- Online News Hour—corporate ethics
- Web-miner—business ethics
- The Center for Public Integrity—investigative journalism in the public interest
- Endgame—online research links

    (See Appendix B for URLs)

Being accountable in the first place is the only remedy against "just deserts." The term dates back to Middle English (its origins from Old French) five hundred years ago—there must be some lesson in that.

Publics and stakeholders

When *your* crisis or disaster happens, you will be on stage before a public audience.

Some of the people you will answer to:

- Families of victims and persons not accounted for
- Your employees and senior managers

- Citizens, neighbors, and the public at large
- Groups, community organizations, trade and professional associations
- Media—local, regional, national, international
- Government—city, provincial/state, federal departments, and politicians

Your phone lines will light up. Journalists will appear with notepads, microphones, and cameras.

Besides everyone else, your own people will want to know what's happening. If they sense something is going on and don't have information, you will be dealing with speculation, rumors, confusion, and, eventually, morale.

Both internal and external audiences will require timely and accurate information. It is important to anticipate your audiences, their questions, and how you would find answers.

Consider different situations and the relevant audiences. Draft your lists. Review them periodically. Gather contact information and background resources. Index these. Date them. Keep them current.

Top-of-mind questions

In a crisis situation, people will be asking you questions and waiting for your response:

- *What happened?*
  The situation may still be developing. It may take on a different shape as new information continues to come in.

- *What is the extent of damage?*
  A physical description and cost estimates are separate issues. One involves simple observation; the other requires full information and a detailed analysis.

- *Who is affected?*
  Persons not accounted for may or may not be casualties. There will be confirmed and suspected victims. Relatives need to be notified first. Before releasing names publicly, review any privacy laws that apply to your jurisdiction.

- *What is being done?*
  Talk about action accomplished and underway, the agencies involved, contingency plans in the event that barriers arise.

- *Why did this happen?*
  The real cause may take time to determine. What seems true today may change in light of new information. Don't assign blame. Don't speculate.

- *When will things return to normal?*
  You may not be in a position to predict this. You can express your best estimate, and qualify any statement, "If everything goes as planned … we will be keeping the public informed."

- *What will be the final outcome?*
  This frames the event in a larger context. It may state a needed change in practice or legislation. The goal is to assess the lasting impact on a community.

- *How are the families reacting?*
  Media will want firsthand accounts from families, rescue workers, and decision makers. You can facilitate families talking to the media on their own terms. Post a list of journalists with their questions and contact information for families to consider. (See Appendix A)

- *When will there be an update?*
  It's important to establish a pattern of information updates, ideally coordinated to accommodate media deadlines.

Where and when

Where will you be when a crisis occurs?

Consider the possibilities:

- A week has 168 hours—less than one in four is a regular working hour.

- You could get a call at 5:00 AM on a Monday, asking whether you can make it in to work in thirty minutes.

- You could arrive home from shopping at 8:30 PM on a Thursday to find the phone ringing. On the line is a journalist asking you to comment on an event that took place twenty minutes ago. He picked it up on an emergency frequency scanner.

The power may be out. Gas pumps may not work. The roads may be impassable. It may be impossible to reach some of your key people.

You may be the point person in a disaster scenario. You can be faced with *two* crises. The disaster itself is one thing. Communicating about it is a separate issue.

You need resources to be at hand. Time is compressed during a crisis. Twenty minutes can seem like hours when you're waiting for information to come in. Hours can seem like minutes when you're totally absorbed in a dozen tasks. Readiness means being prepared for action. The resources you have at hand when a crisis breaks are the ones you will use.

Disasters are crises-done-large. By definition, they involve an entire community. If you are prepared, your organization can play a leadership role in the response.

Recommended reading

Three excellent articles are worth reading and discussing:

- "Critical Infrastructures under Threat: Learning from the Anthrax Scare" (4,500 words) *Journal of Contingencies and Crisis Management*, Volume 11, Number 3, September, 2003. This article discusses the escalating breakdown of critical infrastructures during a crisis, an emerging issue—especially important for CEO, public relations, business continuity, information technology.

- "Lessons Learned or Lessons Forgotten: The Canadian Disaster Experience" (7,400 words) Institute for Catastrophic Loss Reduction, paper by Joe Scanlon, Director, Emergency Preparedness Unit, Carleton University. This paper explores learning, myths, and the reality of disasters—a Canadian example with wider application—especially important for CEO, public relations, business continuity, human resources.

- "The Westray Mine Explosion: An Examination of the Interaction Between the Mine Owner and the Media" (12,000 words) *Canadian Journal of Communication* (online), 21(3), published Masters thesis from Mount Saint Vincent University. This paper is an in-depth analysis of economic, political, and media issues during a major crisis—many useful insights—especially important for CEO, public relations, human resources, department heads.

(See Appendix B for URLs)

◆ ◆ ◆

Scenarios—Questions for discussion:

1. Consider both your organization and its community. What are some scenarios you are likely to face? Is there a scenario that is unlikely but would have grave impact?

2. Do you know which other players you would cooperate with in various situations?

3. Does it make sense to develop resource files on scenarios and related resources?

4. Can you portion this out for people to research through the Internet, libraries, and networking?

5. Which scenarios do you need to prepare for? How would each choice add to your overall resilience?

6. Who will coordinate the monitoring and follow up on this?

3
RESOURCES

There are helpful resources that can be prepared in advance, anticipating critical situations you might face. Some are core items. Others are issue-specific. Together, they give you more time for two critical tasks: dealing with the crisis situation itself and communicating effectively. (IMPORTANT: See also Part 4, "Resilience")

Crisis Readiness Checklist

These ten components need to be in place before a crisis situation occurs. They will enable you to maintain poise and concentrate on your top priority—the crisis response plan.

1. *Public relations policy and procedures*—A statement of mandate (authorization), values (what's important), program (plan of action), leadership (accountable person). You'll know you have these when you can see them on paper. A simple, draft PR and media coordination policy is in the appendix.

   (See Appendix A)

   Also, the communications policy of the Government of Canada includes an excellent section (item #11) on crisis and emergency communications and is available online.

   (See Appendix B for URLs)

2. *Crisis communications action plan*—Key people, roles, action sequences, scenarios. We'll review an action plan a little later.

3. *"Big Picture" information piece on your organization*—A concise history, your mandate, your management team, last year's performance indicators, this year's goals, profiles of major divisions to offer insight into your operations, some basic statistics to help define your organization and its scope. An excellent annual report has most if not all of this.

4. *"Window" information piece on every major program*—This is a one-page document with short paragraphs on each program's or division's roots, mandate, current operations, number of staff, and major roles, a map or list of locations. It must be current and simple. It can be done on special masthead in PDF format. Senior managers can maintain the sheets for their areas. A writer can polish them for conciseness, consistency, and readability. They should be updated regularly. Each item should have a code in the footer that indicates its topic and revision date.

5. *Indexed reference files on potential crisis situations*—This is an issues management exercise. For each potential crisis issue, research and log: minutes, reports, press clippings, focused contact lists.

   Create an annotated index for each issue, indicating whether each item is only a paper file or whether it's a computer document (include format, file size, revision date). Provide the precise location of physical files—building, room, file cabinet, drawer, and folder title. You can develop a code and map for this. Someone who has never pulled the file should easily be able to retrieve it. Test this using a person from a different location.

   The annotated index can work well in HTML format, using links to navigate and drill down following an issue track. This can simply be an HTML folder that's zipped and distributed whenever it's updated. The folder should have the revision date as part of its name—for example: Issues—YYMMDD—that way it will be obvious if everyone has the current version.

If the index is on your intranet, you'll likely want authorized access to various sections on a need-to-know basis. Be sure there are no back-end loopholes through which hackers can access this information. Use a physical firewall plus software to secure unused ports. Be careful to secure any WiFi networks. Change passwords on a regular basis. Passwords should not be obvious (like your address) and should not be kept in open view.

You should also have a folder of hard copies on each important issue, updated monthly and kept in a portable file box. (Paper files are a boon when you need to lay something out before you.) If there is a lot of material, you may want to organize it by year. A few packs of Post-It notes can be very handy to have in the box.

6. *Key person list*—This is a list of your senior management team, with an appendix that shows the senior person and his or her alternate at every physical location. It should include work, cell, and home phone numbers. While you may know your senior team by face, name, and title, everyone may not fully appreciate the background and mandate of each individual.

For every member of your senior management team there should be two facing pages that work together. On the right is a one-page job summary; on the left, a one-page bio showing his or her background and key competencies. Photos can be helpful. At the front of the list is an index.

Trust is invaluable in a crisis situation. Like the oil in an engine, it helps everything run cooler and more smoothly, increases performance, keeps things from seizing up. Your key-person list, as described above, helps foster trust.

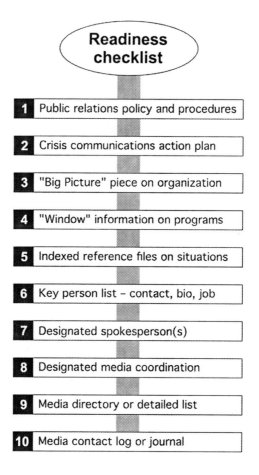

7. *Designated spokespeople*—Default spokespeople should be established prior to a crisis situation. See who is a good public speaker. Some people are naturals at this, others just "lose it" in a stand-up speaking environment. They either freeze or ramble on, not staying on topic or on message. You don't want to learn who's good and who's not in the midst of a crisis. Consider both competence and confidence when choosing your spokespeople. Focus, clarity, and poise are all important.

Your spokespeople and your public relations counsel should know each other. They should preferably have worked together before, even if only in an exercise, so everyone knows what they can expect from the others.

8. *Designated media coordination*—The media coordination function should be established as credible and helpful with both your staff and the media prior to a crisis. Trust is an outstanding asset in the midst of mayhem. If, in the normal course of things, media contact is well coordinated through public relations, then everyone will follow this course in a crisis. Both the media and your staff need to feel this is a natural and effective way to work together. Trust like this is earned, and it needs to be in place prior to a crisis.

(See Appendix A)

9. *Media directory or a detailed list*—You should have a concise list of the major media and your public relations counsel at home with your list of senior managers and spokespeople. Use a commercial list like Gebbie Press, Bowdens, Matthews, or your own contact database. This can help in contacting the media. Equally important, it makes the media more "real" and gives everyone a sense for this important audience.

(See Appendix A)

10. *Media contact log*—You can easily have a dozen or more newspapers and radio and television stations on the go at one time. If an event becomes national news, multiply that number by at least four or five.

Keep a separate tracking sheet for every journalist and story. Know who contacted you, when, about what, how to contact them, what their deadline is, what you promised, whom you've delegated to, when they're due to get back to you, whether you need to follow up. You will *never* remember all this. That's why you write it down.

When you're swamped with calls, hours will seem like minutes. When you're waiting for someone to get back to you, minutes will seem like hours. A simple log can work the way chits do for a short-order cook. You work on your top priorities while scanning the chits for the next priority item. At the end, you have a record of what came in and how everyone handled it.

Information useful to record

Use a one-page form to capture the following information for all media queries:

- Day—date—time

- Journalist—staff/freelance

- Publication/station

- Phone and extension—fax number

- Web site URL—mail address

- Topic—approach to story—who else they're contacting

- Deadline—section/program—date the story will run (if it is a feature)

- Contact and research notes:
  Date | time | who/person | contact/action/follow-through
  (at least a dozen lines with four columns for above)

- Attached*—an asterisk after an item indicates sheet(s) attached

- Page ___ of ___ pages (lets you extend information)

- Initials of the person completing the form

(See Appendix A—also, a PDF file for printing is available at www.topstory.ca)

Emergency e-mail access

Global Roaming enables you to connect to thousands of ISPs in more than a hundred countries—without changing your Internet Service Provider. In most cases you connect by dial-up, which can seem wonderful if the alternative is no Internet access. Your ID and e-mail client remain the same. (Everyone should be familiar with Web-based e-mail in case they need to work from a borrowed computer.)

(See Appendix B for URLs)

Web posting kit

The Internet is the first place many people will go for information and breaking news.

Planning the following resources will help you effectively use the Internet:

- A link from your home page to a crisis update page

- A crisis update page with the latest information in reverse chronological order (this is a simple and direct way for people to get the latest and recent news)

- A media page, either unindexed or requiring password access (you can provide the URL to the media in advisories and news releases—high resolution files you may offer to the media will put a heavy load on your bandwidth if everyone can download them)

- A means to archive the content of crisis and media pages as information ages (as content accumulates, you can trim off outdated material and repackage this as archives, either as Web pages or linked PDF files)

You need to be able to post new information to your Web site even when personnel who normally do this are not available. There are three ways to do this:

- A content management system (CMS)

- HTML posted by cross-trained staff
- Outsourcing or reciprocal agreements with peer organizations

Content Management System (CMS)

A CMS system enables you to manage content from anywhere, as long as you have an Internet connection and a standard Web browser. You go to your site, enter a username and password, and editing tools appear on pages that are CMS enabled. You make the additions or changes you want, and then log off. The instant you log off, your changes are live on the Web site.

If you're not familiar with CMS, WebYep is a simple system that offers an online demo:

(See Appendix B for URLs)

If you can edit three pages on your Web site: "Home page," "Crisis updates," and "Media resources," you can manage. You can offer archived content as downloadable PDFs.

There are many open source CMS systems. Your webmaster can review these and discuss content management with you. If you don't already have a CMS system, start simple.

Content management system resources:

(See Appendix B for URLs)

Having CMS is not enough. You should implement a realistic and fully developed crisis information page as part of your testing drills. This can be done on a local server so it doesn't disrupt your public Web site. You won't know what barriers you'll face until you've met them.

Without CMS, you will need to create and post HTML pages to your Web server. Even if you have a webmaster, *at least one* other person on your team should be cross-trained in this task. This person should participate in your crisis drills, using a desktop folder that contains your complete site. For FTP practice, this person can post some regular pages to the public server.

HTML kit for cross-trained staff

An HTML kit with the following items will help you meet important deadlines:

- Home page—your home page with a new link to your crisis update page

- Crisis update page—a template page with a placeholder for your information—for access to dated content, you can use links to PDF files

- Masthead—word processing template for repackaging dated content as PDF archives

- Media resources—another template page (to reduce other, non-media, traffic, you may want to set your robots meta tag and robots.txt file for search engines to ignore this—you'll give editors and journalists the URL in news releases)

- If your site uses Google Maps, instructions to update the refresh date for a page in the sitemap.xml file

- Clear and complete printed instructions with phone numbers for support

- Full server access information (FTP address, username, password)

- Full support information (phone and e-mail address for your hosting company's support)

- A good HTML handbook for easy reference, such as: *HTML 4 Unleashed* by Rick Darnell (ISBN 0-672-31763-X)

- A CD with your complete Web site folder, resources, and installers for HTML and FTP programs—including freeware—can eliminate any machine-specific licensing issues (you can download this right now—see below—and save it in a new folder)

If your support person is without his or her regular computer, this freeware may be useful:

- Windows freeware HTML editor—Araneae

- Windows freeware FTP client—Coffee Cup Free FTP

- Mac OSX freeware HTML editor—Taco HTML

- Mac OSX freeware FTP client—CyberDuck FTP

- Simple, clear HTML examples—W3Schools

- Excellent UK site with beginner/intermediate/advanced guides—HTML Dog

   (See Appendix B for URLs)

Important points:

- Designated support persons should have hands-on experience working to deadline.

- There should be instructions and a decision tree for configuring, resetting, and troubleshooting modems and routers—prove the instructions in a testing exercise.

- You should include dial-up practice in case there is a problem with DSL or cable.

- If your regular ISP is not available, a global roaming resource can save the day. (It is important to set this up in advance—there are no charges unless you actually use the service.)

- Your hosting company should have the names of people authorized to work on your account during an emergency—usage may require an authentication phone call.

(See Appendix B for URLs)

In a crisis situation you may want to post updates 24/7. Your regular support staff may be unavailable due to illness or evacuation. Consider how many people you want to cross train. The Web is a *critical* information channel, and you need to plan for contingencies.

Reciprocal agreements and outsourcing

There may be a peer organization, ideally in another region, with which you can develop a reciprocal agreement. Either of you can call on the other if your Web support staff is not available during a crisis. You can negotiate an hourly rate for emergency support. There may even be a small network of three or four organizations interested in such an arrangement. You should have a *formal* nondisclosure agreement, signed by everyone involved.

You can work together on the parameters for a support kit, then make your server account and log-in information available to each other in an encrypted file. During an emergency, the supporting organization gets the password for this encrypted file. As soon as the emergency is over, you can change the password on your server account. If you are a large organization, your IT department can set permissions to restrict access by emergency support to specified areas.

Another option is to outsource emergency support for Web site work. The advice regarding a different region or city and a signed nondisclosure agreement holds here as well. You may want to consider some sort of retainer agreement to ensure availability.

Really simple syndication (RSS) feeds

RSS feeds for your Web site are worth looking into (RSS stands alternately for: Really Simple Syndication, Rich Site Summary, RDF Site Summary). A growing number of people use RSS newsreaders—software applications that list the headlines of subscribed-to RSS feeds and, with a click, take you right to the content. People who subscribe to your RSS feed are alerted whenever new information appears on the page that holds the feed.

Some online resources:

- MarketingSherpa—Practical news and case studies—RSS feeds
- Wikipedia—RSS (file format)
- Introduction to RSS
- "How to feed RSS—a hands-on guide"

(See Appendix B for URLs)

Portability—USB flash drives

A portable USB flash drive (the size of your thumb) is an easy and inexpensive way to *always* have your crisis resources with you. Thumb drives are available in sizes from 32 MB to more than 4 GB capacities. You copy information to one just like transferring it to another hard drive—you can transfer 500 MB in less than a minute. (Your crisis resources, including your full Web kit, will be far smaller than this.)

In a crisis situation, if you're not at work, you just insert the USB flash drive in any computer that's available, and your resource files are right there.

Any memory device should be in regular use, so you know it's functioning. There is a *mandatory* sequence for removing USB flash drives. You must first unmount (eject, trash) the USB drive icon, and then wait for the USB drive's indicator light to stop flashing. When the icon has disappeared from your desktop *and* the flashing has stopped, you can safely remove the drive. Not following this sequence can render the contents useless. Read the instructions that came with your USB drive, including any precautions regarding care.

If security is an issue, some manufacturers offer models with biometric keys or hardware encryption. (The Kanguru Micro Drive AES, for example, meets U.S. federal requirements for securing sensitive data.)

(See Appendix B for URLs)

Portability—Hard copy files

The power may be out or you may not have access to a computer. Keep a hard copy of key resources, indexed and portable, both at your office and at home. A portable file box about the size of a phone book can hold 200 pages of key information. If you're using binders, it's useful to organize files into several half-inch ones. That way you can have briefing notes, media lists, and your key person directory all open at the same time—each binder works like another computer window. If you have extensive briefing notes, organize them by topic into several half-inch binders with thin, flexible covers.

Portability—PDAs

PDAs do these things well:

- Keep full contact information
- Enable precise time/task tracking
- Give you access to brief documents
- Let you beam files between devices
- Manage appointments and reminders

- With a modem—access e-mail and send news releases

It's easy to keep full resources on a PDA, such as a Blackberry or a Palm device. Software such as Documents-to-Go for Palm devices and eOffice for Blackberries, along with Secure Digital (SD) cards, makes this easy. However, it's difficult to transfer information to a desktop computer unless it's your own, and PDA screen size makes it impractical to view long documents. A PDA is most useful for contact lists. During a power outage, it's important to have a car charger for your PDA.

A little-known Palm device that can be very useful during an extended blackout is the Dana. It has a wide screen, a full size keyboard, two SD card slots, and runs for twenty-five hours on either its built-in rechargeable battery or three AAs. Built for the education market, the Dana is designed to withstand a four-foot drop. Using Documents-to-Go, it's easy to sync your Dana with a laptop, transferring Microsoft Word and Excel files between the two.

Palm users can back up all their data to an SD card using BackupBuddy VFS. If your battery goes flat and you lose all your data, recharge the battery and click Restore—you'll be back in business in less than ten minutes. You can routinely back up to two SD cards and keep one in a safe place. Then if you lose your device or it is stolen, insert the SD card in a new Palm and click Restore—all your applications and data will be installed. SD cards are the size of a postage stamp, thin, and inexpensive.

If data security is an issue, TealPoint offers TealLock, a security and data encryption solution for Palm devices. It has a whole range of options, including the ability to shred all data if the wrong password is entered a specified number of times. There's also a corporate edition that offers multiunit installation. BlueFire Technologies offers Mobile Security Suite for Palm and Windows Mobile PDAs—it can remotely erase all the data on a device. Blackberry Security offers protection for both wireless and stored data.

(See Appendix B for URLs)

Portability—Cell phones

It's a good practice to have cell phones served by two independent networks at every location. You can poll your staff to see who each person's carrier is. (Periodic surveys will help keep this current.) Then if one network is down, you will have the name of another person in the same department who can be reached.

Maintain this information in an Excel spreadsheet or workbook with columns for name, cell phone number, department, location, and network. You can then do a multiple sort by column—network, location, department, name, and number.

With many newer cell phones, you can send an SMS message (small message service—there's a size limit) to a group with one call. If you have a crisis notification message saved as a template, you can quickly alert your entire senior management team with one SMS message, then follow up with individual calls. Test this to see how the message comes through and how people respond to it.

With many carriers you don't need to subscribe to SMS in order to receive text messages. And you can even send them—each message just costs a few cents more.

It's a good idea to have a short list of key media contacts on your cell phone. At the very least this should include major radio, television, and newspaper editors for your city, plus leading national media, key trade publications, and newswire services. With these numbers in your phone you can still reach someone if your cell phone is all you have.

Many current cell phones can hold two hundred to five hundred contacts, with multiple phone numbers for each. Bluetooth or infrared connectivity is important because they enable you to quickly upload detailed lists as

VCF (virtual card format) files. Even if this needs to be done one contact at a time, it requires only seconds per record. One computer or PDA with a Bluetooth port or an infrared port can transfer a list of contacts to multiple cell phones, if they have the corresponding port.

If a cell phone with sensitive contact data is lost or stolen, some networks (Sprint, for one) offer a kill service that erases all data on a cell phone by sending a special signal to it. For the delete instruction to work, the cell phone just needs to be turned on. If you think this might be important, check with your provider to see if they offer this service.

Languages and culture

The United States is a melting pot; Canada is a mosaic. More than a million people migrate to North America each year. Immigrants represent more than 10 percent of the population. The native tongue of many North Americans is a foreign language.

Even if people know English, in a disaster situation it's easier and more comforting if someone speaks to them in their native tongue. Many hospitals have a list that cross-references their staff with languages understood or spoken—it's not uncommon (in the largest cities) for sixty languages to be represented. That way, in an emergency, someone is available who can assist with communication.

Understanding both what people are saying and their culture is important.

It's simple to build a list of multilingual people in your organization. Include person, location, and language. A onetime survey can bring you up to speed on this. You can then adjust the list as people come or go.

Whether or not language is an issue, culture is important. People tend to take their own culture for granted. Consider what this means when dealing with a culturally diverse audience. Everyone interprets messages and behavior from his or her own cultural perspective.

Your staff can give you a sense for what is important to them because of their own cultural background. To look into this further, there may be cultural associations in your area. Some excellent resources on cross-cultural sensitivity and communication are available online:

- Foreign Affairs Canada: Centre for Intercultural Learning
- Internet Public Library: Ethnicity, Culture, and Race

   (See Appendix B for URLs)

Language or culture may become a communications issue during a crisis or a disaster. You can't predict which specific resources might be relevant. But you can document the languages spoken by your staff and familiarize yourself with community and online resources.

The time to consider and prepare resources is *before* the event. Readiness takes time.

◆ ◆ ◆

Resources—Questions for discussion:

1. Which of these resources do you already have on hand?
2. Are there any you would add to this list?
3. How would you do an inventory and prioritize the outstanding items?
4. Who would be responsible for producing the items you still need?
5. What would be your ideal budget and timeframe?
6. If you don't have sufficient resources, could you spread the cost over time?

7. Would items be charged to general administration or to individual department budgets?

8. Which departments are prepared to allocate resources to this?

9. How would you integrate crisis response resources into your regular budget planning?

4
ROLES

If you don't step back and examine roles, you will never be aware of presuppositions and assumptions—both yours and those of others.

Crises and disasters involve multiple players. Each of them has a specific role.

Aside from your own organization, three levels of government—municipal, state or provincial, and federal—can be involved in any disaster. One level will have primary responsibility. However, involvement can quickly escalate to higher levels. In Canada, for example, municipal government is the first level of response. But any event involving a nuclear generating plant immediately involves the provincial government.

Clarifying roles

One of the first things to do is to clarify roles and assess the accountability and performance of your organization. This serves two purposes: It can help you track and manage issues to prevent or mitigate a crisis or disaster in the first place. It also helps define limits for your accountability and response.

A public health example

An employee becomes sick and collapses at work. You call 911 and the person is taken by ambulance to hospital. He had taken public transit to work, and his job involves public contact. He is diagnosed as having a highly contagious and potentially fatal disease. Public health becomes an immediate issue.

Follow the steps in this course of accountability:

1. As an *employer*, your responsibility is to be responsive to the person's condition and get him to medical care as soon as possible.
2. Once you've dialed 911, responsibility is passed to the *ambulance service* to transport the person quickly to medical care.
3. Once the person arrives at *hospital*, that institution is responsible for diagnosis and care.
4. Once the hospital diagnoses a disease such as SARS or meningitis, *public health* joins the team and deals with etymology, risk assessment, interviewing the bus driver, tracing (with your help) customers who were on your premises that day, etc.

Four agencies are involved—employer, ambulance service, hospital, and public health. Each plays a specific role. If the situation became public and the media were involved, specific questions would be directed to each agency according to their role.

Coordinating roles

In this example you as employer, together with the ambulance service, the hospital, and public health are a de facto incident response team. You would talk and decide who should field which issues.

All of you would likely prepare media releases and share them with each other. The releases might not be used unless the media initiated contact. Public health would likely be the dominant influence, wanting to avoid uncalled-for worry or panic.

For your part, you could confirm that an employee had become ill at work and was taken to hospital.

The ambulance service would deal with questions regarding response time.

The hospital would handle questions regarding diagnosis and the patient's condition. They will not release the person's name without proper consent. Both you and the hospital would refer any questions regarding contagiousness and risk to the public health team.

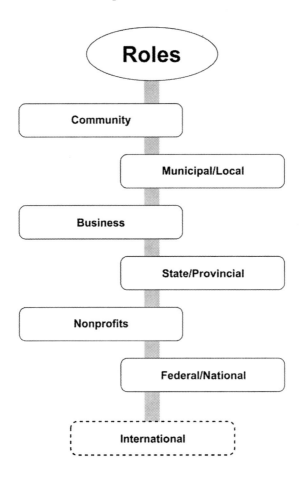

Public health would interview your staff, the bus driver, and the patient's family. They would trace who the person was in contact with and advise people about vaccines, symptoms to watch for, what to tell their family doctor. People they speak to would likely be given an information sheet and a phone number to call if symptoms develop.

Fielding questions

Your employees would likely be asking about what happened, the person's condition, whether they or their families are at risk. Clarifying and coordinating the roles of everyone involved will help you answer these questions.

You could tell employees what the hospital said about the patient's condition. You could assure them that public health is involved and is following through to assess and manage any health risk.

If the media became involved, you would answer questions appropriate to your role. For questions involving the role of other agencies, you would refer journalists to them as the best source. This works well, as long as respective roles have been clarified and confirmed.

This public health example is a simple illustration, but the principle involved applies across the board. It's good to include complex situations, and actively involve multiple agencies in your planning and testing. Getting to know one another establishes valuable contacts, relationships, and trust.

Teamwork

Responding to a crisis involves more than one person. All those involved need to know what position they play and whom they can rely on for what when it's their turn to pitch in.

IMS model

Governments use an incident management system (IMS) model in which the top level typically involves five to eight people. IMS models have evolved from extensive experience with disaster situations.

You need to determine who is on your core team for various situations. If your organization has three or four management levels, here is a suggested minimum:

- *Chief executive officer*
  Brings all resources to bear on the crisis and its resolution

- *Public relations manager*
  Leads the communication process, coordinates media contact

- *VP/department head involved*
  Leads his department in responding to crisis

- *Facilities/location manager involved*
  Involved hands-on; becomes best direct information source

One or more of the following could supplement these:

- *Human resources manager*
  Finds appropriate support staff, handles employee relations

- *Chief financial officer*
  Finds funds for special resources

- *Office services manager*
  Provides administrative support, information files, phones, and reception

- *Legal counsel*
  Provides advice on potential legal issues and liabilities

You should have a crisis action plan with *timelines* and *roles* that you've developed as a team and tested and refined in regular exercises.

Also consider outside agencies you will cooperate with in different scenarios. You may want to involve them occasionally in joint exercises.

The incident management system is an important model to know. All levels of government in Canada and the United States have formally mandated it. All government bodies in North America are familiar with it.

A typical IMS has five functions:

- *Command*—Liaising with other agencies, establishing communication strategy, and media relations; monitoring action; managing resources; assuring overall safety of staff
- *Planning*—Gathering information, assessing the situation and contingencies, creating an action plan that identifies objectives and incident response activities
- *Logistics*—Providing facilities, services, materials, personnel to respond to the incident; organizing and verifying availability of staff
- *Operations*—Directing and coordinating operations, verifying safety, assisting in developing and implementing the action plan, requesting or releasing resources, providing continuous updates on implementation of the action plan
- *Administration*—Tracking all expenses, claims, purchases, and contracts initiated during the emergency; identifying existing resources that were depleted during the incident

You need to review these functions as a team and develop a consensus regarding who will handle what. The onslaught of a disaster is not the time to begin this process.

Useful sources for information on IMS:

- FEMA: National Incident Management System

- Environment Canada: Environmental Emergencies

  (See Appendix B for URLs)

Software relating to IMS:

- Incident Commander
- Incident Management System

  (See Appendix B for URLs)

Teams

Responding to a crisis involves teams. A community response involves teams of teams.

Katzenbach and Smith have written an excellent book—*The Wisdom of Teams*—that examines what makes teams work:

> "A team is a small number of people with complimentary skills who are committed to a common purpose, performance goals, and approach for which they hold themselves mutually accountable."

The authors cite six attributes that are common to successful teams:

- *Small enough*—large numbers of people have trouble interacting effectively as a group; challenging subsets of the group to tackle specific performance goals and then helping these subgroups become real teams is a better way to involve many people.
- *Complimentary skills*—three kinds of skills are important: technical or functional expertise, problem solving and decision-making skills, and interpersonal skills. No team can achieve its purpose without all the skills required.
- *Truly meaningful purpose*—this provides meaning that guides what the team is to do. A team's near-term performance goals must always relate to its overall purpose.

- *Specific goals*—these are compelling and provide clear and tangible footholds; small wins build commitment.

- *A clear working approach*—agreeing on the specifics of work and how it fits together to integrate individual skills and advance team performance is important.

- *A sense that "only the team can fail"*—unless the group holds itself accountable, there can be no team; a team is a group that is truly committed and jointly accountable for results.

The *Wisdom of Teams* has a much wider application, but you may find this book useful in thinking about and developing your crisis response team.

◆ ◆ ◆

Roles—Questions for discussion:

1. How would you determine your accountability in a crisis situation?

2. What are the agencies you would likely collaborate with in various scenarios?

3. Who would be on your crisis response team for two very different scenarios?

4. How would you assign the functions of an incident management system (IMS)?

5. Is there a team building exercise that would help your people to clarify their roles?

6. Have you developed contacts with other responding agencies?

7. Are there joint team building exercises you can do with outside agencies?

◆   ◆   ◆

ENDNOTE:

The Teams section of this chapter is adapted from Katzenbach, Jon R. and Smith, Douglas K. *The Wisdom of Teams: Creating the high performance organization.* New York, New York: HarperBusiness, 2003 and used with permission of the authors.

5
PROCESS

This is the critical path for managing crisis communications.

Hospitals have a code system for indicating an emergency. Many jurisdictions are adopting a uniform system. One common system uses *code blue* for cardiac arrest, *code red* for fire, *code black* for bomb threat, *code orange* for external disaster, and so on.

When a code is called, hospital staffs know immediately what is happening, what the priorities are, and what each member on the response team needs to do. You'll see staff abruptly stop what they're doing and walk briskly to their stations. It's worth their jobs.

Crisis code

While it's unlikely you'll need a system as detailed as those used in hospitals, you must have a way to clearly and quickly communicate that your organization is signaling a crisis response.

If a crisis is unfolding, you don't want your secretary to wait until you return from lunch or a two-hour meeting to inform you. You should have a simple code that everyone understands. It should tell them that there's an emergency and an urgent response is number one priority. A code is the "start button" for your crisis response process.

Something as simple as "code C," with a simple and clear hand signal if someone needs to be called out of a meeting, will do.

There should be metrics or timelines to ensure a proper sense of urgency. For example, when a crisis occurs, one of three specific senior managers must be notified within five minutes. Your full incident management team must be notified within the next ten minutes.

You should test this on a regular basis, including evenings and weekends. If a code is called during lunch hour and a senior manager is at a restaurant, someone should phone and have him paged—or go there and find him. The incident can't wait an hour. You need to walk through this process so that it will work when a real crisis occurs. In a fast-breaking situation, every minute can make a difference. This is a test of your corporate culture and its sense of pressing importance, its sense of urgency.

Media queries

It might take you a full day or longer to do the work you'd like to put into a news release. You won't have that time. Newspapers, radio, and television have deadlines. If media calls are flooding your phone lines, you don't have hours to respond. If you wait too long, you will be labeled as unresponsive or as someone who doesn't return calls.

You should try to have an initial media release ready within one hour, whether you will use it right away or not. It may just acknowledge the crisis, tell people what action you are taking and that you will be issuing updates as more information becomes available. Once this is prepared and "in the can" you should aim for a more detailed release within the next two to four hours, as soon as you have more reliable information.

Here is an outline for an action plan that generates an initial response. Once you've issued a release, the process will repeat itself and develop as new information comes in and action to deal with the crisis progresses.

Initial response

Initial response—action plan outline (in minutes):

- *Notification*—00:00 to 00:15—report up, call crisis code, notify full team

- *Research*—00:05 to 00:60—gather information (full team) (concurrent)

- *Analysis*—00:15 to 00:60—assess and plan (full team) (concurrent)

- *Communication*—00:30 to 00:60—draft news release within one hour (concurrent)

- *Evaluation*—ongoing (also pull everyone together for a final evaluation within one week of the crisis, before impressions start to fade)

All five points form a process cycle that repeats itself and flows throughout a crisis. You are always being notified of developments, gathering additional information, considering next actions, updating people, and taking the measure of things. A crisis is a fast-paced and turbulent environment.

Timeliness is important. When something is demonstrated to be immediately apparent, this supports its credibility. Misinformation sets like a stain and needs to be blotted up immediately. Crises demand a sense of urgency. You should be ready to deal with the media within an hour—two hours at the most—of the very start of a crisis.

Acknowledge all media calls within twenty minutes. If a journalist asks for information you don't have, offer to get back to her and outline the process. You may need to speak to a department head, who then contacts the right supervisor, who then finds someone on the scene; then the information needs to get back to you. Appreciate deadlines and be helpful. State the importance of accuracy and be realistic.

If all your resources are committed to gathering information, you may need to state this and point journalists to your upcoming news conference or to media resources posted on your Web site. Nevertheless, always try to help whenever you can.

Trust is paramount in a crisis situation. The public expects strong leaders who are competent, emotionally intelligent, and in charge of the situation. A timely and appropriate response sets the tone for media coverage in the days to come.

Here is the initial response action plan in more detail:

Notification

Everyone must know that your senior team requires *immediate* notification when a crisis occurs. Medical people have a word for this: Stat. It comes from the Latin adverb *statim*, which means without delay.

Notification—00:00 to 00:15—(local):

- The absolute first priority is that people involved—next-of-kin, neighbors, etc.—are being dealt with in the most sensitive and appropriate way.

- Staff involved immediately inform their department head, public relations, and the chief executive officer. If one person is unavailable, they should immediately proceed to notify the next. Once one person on your crisis team is reached, that person can ensure the full team is notified.

- Public relations immediately advises the crisis location and the central switchboard regarding how to handle media calls.

- The chief executive officer contacts appropriate senior managers and sees that the board executive is notified of the situation.

Research

Accurate information, often from a number of sources, is essential for considering your response. Several people can work on this simultaneously. All information should be tied to sources.

Research—00:05 to 00:60 (department head, CEO, public relations)

- Speak with staff at the crisis location and gather information.
  (department head leads)

- Consult with responding agencies such as police, fire, public health, hospital.
  (CEO leads with support from public relations)

- Review issues management material on file and speak to reporting staff.
  (CEO leads with support from public relations and department head)

- Monitor and record radio and television news; check media Web sites for breaking news.
  (public relations leads with support from administrative services)

- Take detailed notes. It's important to write down point-form notes.
  (log the time and source for all incoming information)

Analysis

Information needs to be viewed in a number of contexts to determine your approach and next steps. Ideally, your team will have a consensus on important items.

Analysis—00:15 to 00:60

- Assess the responsibility and performance of your organization.
  (CEO leads with support from department head and public relations)

- Identify the key publics involved and their communication needs; check this with location staff.
  (public relations leads with support from CEO and department head)

- Review issues management, media response, where editors and journalists may be coming from, the tie-in with broader issues.
(public relations leads, confirms with CEO and department head)
- Select your approach, the key points of your story, your communications strategy.
(public relations leads, confirms with CEO and department head)

Communication

You must prepare an initial media release and brief relevant staff within a very short timeframe; ideally within one hour. You can refine this release as new information comes in or while waiting for media calls.

Communication—00:30 to 00:60 (for initial media readiness)

- Draft an initial news release within the first hour of crisis notification.
(public relations, in consultation with crisis location, CEO, and department head)

- Brief staff at crisis site and potential spokespeople regarding approach, story points, rationale, communications strategy. (Within twenty minutes of completion of the news release—this can be a three-way call with your spokesperson and the lead person at the crisis site.)
(public relations, in consultation with crisis location, CEO, and department head)

- Direct all media calls to public relations.

- Public relations acts as an information hub: logs calls, refers journalists to appropriate sources, coaches spokespeople, sends out media kits, arranges news conferences if required, keeps CEO informed. (Crisis event may be brief or may last days or weeks.)

- Your organization projects competence, integrity, and sensitivity.

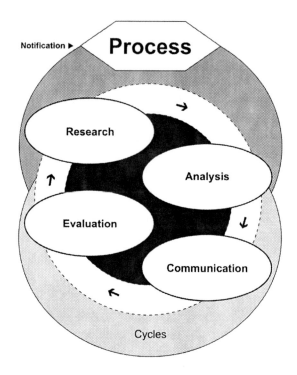

Evaluation

Appraising the appropriateness of your response will help direct future action.

Evaluation—within three to five days after the crisis ends:

- Assess response soon after the crisis has passed, otherwise impressions will evaporate.

- Contact key people and ask, "Now that we've been through this once, how can we improve our response the next time?" Ask everyone to file free-form "hindsight" notes. Make it clear that you are evaluating the process, not people.

- Record response times; establish norms and goals. Crises demand a sense of urgency.

- Rate trust. People will know how they feel about this.

Breaking news

When releasing news, keep these points in mind.

Excellent internal communication is critical to effective crisis communications:

- You should have a current "need to know" list regarding various scenarios.

- The first people to know of a crisis will be immediate responders. Information needs to pass up and across your organization in a timely fashion. Post all releases internally the minute they are issued to the media. Ask managers to mention postings.

- Besides your employees, be prepared to communicate with important stakeholders such as investors, regulatory bodies, industry and professional associations, and your supply chain both upstream and downstream. Develop a grid of departments and the stakeholders they normally deal with and another grid of stakeholders versus issues.

A media release should be carefully considered regarding the public interest:

- An urgent situation requiring action to ensure public safety should be the first test. If people need to protect themselves, or if public assistance is required, the need for publicity is obvious. It's also important to be transparent where public accountability is involved.

- In the balance is the issue of unnecessary fear or panic. A classic example is shouting, "Fire!" in a crowded theatre—a better alternative is to turn the lights on, take control, and guide the audience out in a way that doesn't foster panic. Consider the communication options. Issuing a release should not be a knee-jerk reaction.

- In some circumstances, usually involving a contagious disease, controlled communication with groups at risk may be the way to go. In such a situation, you would be in touch with public health and they would take the lead on this.

- In any situation involving public risk, you want to create the right sense of urgency and channel emotions into appropriate action. There are situations where it's important to raise an alarm. But the *purpose* is for people to take timely and appropriate action.

Practice and functionality

People learn by doing. If you walk through your plan on a regular basis, it will become internalized—it will come to mind naturally when everyone needs it. Chapter 12 deals with testing in detail.

There is such a thing as overplanning. Your action plan should be something people can keep in their heads in the midst of havoc. No one has time to begin digging through a two-inch binder, reading procedures, in the middle of a crisis. Everyone needs to know *The Plan*, and it has to be simple (and therefore flexible) enough for that to be possible.

Communication plan elements

As a backdrop to crisis communications, it's useful to look at the basic elements of any communication plan:

- *Research*—Ready-fire-aim doesn't work. Gather the facts and carefully consider the issues, the stakeholders, and appropriate action.

- *Analysis*—What information do people need or want? What are your meta-messages (major themes)? What are your specific talking points?

- *Communication*—Who are your publics, stakeholders, and major audiences? Do you have current contact information for each of them? What are the options for reaching each audience? What is their natural or preferred channel of communication? Consider every audience and see whether there's a tailored message that would speak directly to them

and their issues. It may be worthwhile to segment and coordinate messages.
(An example: During a major storm, a Jewish area of a Canadian city with many elderly residents needed to be evacuated in the middle of the night. Instead of soldiers in uniform, people dressed in civilian clothes and accompanied by Hebrew translators were sent to knock on doors.)

- *Implementation*—Who does what and when? How long does it take? What is the backup plan if you run into obstacles? Are people empowered to be flexible in taking action?

- *Evaluation*—How will you determine whether communication is effective? If your plan works, what results would you expect to see? What would you expect if it weren't working?

Do you have a written crisis communication plan? If you asked the key players on your team, what would be their answers? How recently have you tested your plan?

♦ ♦ ♦

Process—Questions for discussion:

1. Do you have a clear system for signaling a crisis situation?

2. Is the system for signaling common knowledge among your responders?

3. Do you have team lists for responding to different types of crises?

4. Do you have an action tree that prescribes how notification and response should spread throughout your organization?

5. Have you walked through the process to see whether phone trees work and how long it takes for everyone to be notified?

6. Is everyone on the same page and comfortable with the process?

7. Do you have a list of obstacles you might face in an actual crisis situation? (Remember Murphy's Law.)

8. Are people empowered to improvise if someone drops the ball?

9. Have you done a practice exercise?

10. Do you log your exercises and compare the metrics—what did or didn't happen and the specifics of how long everything took?

6
PRINCIPLES

There are values and standards against which you and others will measure your actions.

Principles to consider

- *Mandate*—How does this relate to your organization's role, authority, and responsibilities? Who else is involved? What are the public expectations? What are the expectations of any governing bodies or peer organizations?

- *Standard of care*—Has your organization exercised due diligence? Have you acted honestly and in good faith? Have you exercised the care, diligence, and skill that a reasonably prudent person would use in comparable circumstances? Have you complied with any acts, articles, bylaws, and agreements? How do your own people feel about the standard of care? Are they comfortable that they have done their very best?

- *Good faith*—Would your actions be generally seen as honest, sincere, and as having no intent to deceive? Even if you *have* acted in good faith, is there room for the perception that you have not? Are you confident you could deal effectively with a devil's advocate?

- *Truth*—Does your information fit with the facts? Is it accurate? How many versions of the facts are there? Are you being rigorous and objective in selecting the version you choose to believe? Are your choices consistent, or do you fly a flag of convenience, choosing only sources that fit with your beliefs?

- *Transparency*—Is everything evident, obvious, and do you act with no attempt to hide or conceal information? Do you readily cooperate with requests for public information? Do you give people immediate feedback and an estimated time if it will take a while to supply information?

- *Timeliness*—Have your actions been timely in a way that's appropriate to the situation? A crisis is an urgent situation and immediate response is a public expectation. Do all of your people understand and appreciate the urgency issue? Is this part of your organization's culture?

- *Emotional intelligence*—Do you show empathy, understanding, and acceptance of other people's feelings? Have you acknowledged or validated the feelings of victims, responders, and their families? Does your staff demonstrate respect for others? Your public face is a mosaic formed by *all* your people. The media will be quick to pick up even one callous comment.

- *No speculation*—Have you made statements without firm evidence and thus risked losing credibility and trust? Do you log the source and time of all important incoming information? Do you attempt to verify important information with another credible source? Do you seek top-level authorization before releasing important information?

- *Privacy*—People have a basic right to privacy, to not be disturbed or observed by others. This goes to the very heart of respect for the individual. Are you familiar with federal and state or provincial privacy laws? Do you have hard copies of any relevant legislation available for easy reference? Do you have examples of any exceptions to the rule, accompanied by the rationale and the principles involved? (There may be exceptional cases where the public interest overwhelms a person's right to privacy.) Have you outlined the process for making an exception and confirmed this with everyone involved?

- *Liability*—Is there anything to which you are held responsible by law? How much room is there for interpretation, and what are the near and far limits? Are there areas for which you're likely to be held morally responsible, even though the law does not hold you accountable? Public expectations hold everyone accountable for 'doing the right thing.'

◆ ◆ ◆

◆   ◆   ◆

Principles—Questions for discussion:

1. Could you review a history of disasters to look for hindsight evaluations?

2. Do you have your own list of standards and considerations?

3. Have you done a team brainstorming session to develop your list of standards and considerations?

4. Have you documented your organization's principles as a values statement?

5. What is the test to determine whether a principle is being met?

6. Do you have reference documents, expert resources, and peer advisors listed for every important principle?

PART 3
ISSUES AND ACTION

Issues management both precedes and follows a crisis situation. If business as usual is a road, then crises are potholes, washouts, detours, and speed bumps. Good issues management compares to a good road maintenance program and safe driving skills.

7
ISSUES MANAGEMENT

For excellent public relations you must manage your interdependence with the community. A good issues management system identifies what matters to people, and then tracks information for intelligent decision making that fosters rewarding mutual relationships.

Tracking issues

Issues management needs to be proactive. It should monitor your environment for emerging concerns, initiate conversations with stakeholders, and negotiate mutual solutions.

As an overview document, you should have a concise table with fields to summarize:

- Issue

- Current status

- Accountability leader

- Priority

This overview should fit on one landscape page. The next level down should have a summary on each issue including:

- Briefing notes by the accountability leader

- An incident chart showing department, location, date, time

- An index or annotated list of reports, minutes, media coverage, etc.

- A list of stakeholder groups with the status of their contact information

You should have up-to-date contact information for stakeholders tied to each issue. If you don't have a list, you should know whether producing one requires original research or if one can be purchased (make sure you know the source, format, cost, and delivery time).

Media audit

It's often worthwhile to do a media audit on important issues. There are news article databases available online, through public libraries, and for purchase. (Some major public libraries allow you to log in remotely and use your library card number to access their proprietary databases.) A search for relevant keywords will show you stories that have been published on a topic, the publications that ran them, and often a short summary. Some databases enable you access to the full text of the article.

It's useful to capture the following information on key issues:

- Database used

- Keywords used

- Publication cited

- Issue date

- Section/page

- Headline

- Byline

- Sources cited—person, organization, role

- Salient points attributed to each source

If you keep this in tab-delimited format (such as a Microsoft Word table, an Excel spreadsheet, or in database format) it will be easy to sort and view the information in various ways.

Salient information

For every person who reads a full story, there will be ten who just scan the salient bits. It's often useful to capture additional information on a critical issue.

Salient bits:

- Headline
- Lead sentence or paragraph
- Bold subheads
- Call-outs (bits of larger or boxed text)
- A note on any photos or graphics used
- Final sentence or paragraph
- Sidebar headlines
- Name, position, and organization of sources quoted

This list comprises the impression someone might gather by just scanning a story. People commonly talk about something they've "seen" in a publication, when they haven't read the full article. Scanning a story is not the same as reading it. The overall impression that's left may well be different.

This is background research and can give you a good sense for the history of an issue, who the major players are, how they are performing, and how this ties in with the public interest. You will have more context for managing issues within your own organization.

Dismantling even a few stories this way will give you a new perspective on how impressions are formed and the importance of issues management.

Periodical index

Two periodical indices offer a fertile way to learn how an issue was covered in mainstream magazines:

- *Readers' Guide to Periodicals*
- *Canadian Periodical Index*

These are huge print volumes available at major libraries. Each book covers about 400 publications, indexed and cross-referenced by subject and author. The *Canadian Periodical* Web site gives an example of how a person quickly found three hundred articles on SARS. If you photocopy a subject listing, you'll have some idea of who covered a story and how it played. You can use this information to follow up on individual articles. Some libraries carry the electronic version of these databases.

(See Appendix B for URLs)

Internet search

Two things to remember:

- The Internet should be only a supplement to other research
- Meta search engines offer much broader results than Google

Going beyond Google ... explore these two powerful options:

- Clusty (formerly Vivisimo) uses major search engines and clusters results (note the advanced and preferences menus)
- Dogpile searches Google, Yahoo, Ask, MSN (note the advanced search option)

(See Appendix B for URLs)

Blog search engines provide a window to online buzz:

- Technorati lets you create a watch list
- Blog-maniac offers an excellent orientation to blogs and blogging

   (See Appendix B for URLs)

Dedicated search software

Two dedicated search applications, Copernic for Windows and DEVONagent for Mac OSX, do extensive online research using specialized portals, and then aggregate the results. Both work best with a high-speed connection, and it may be useful to do a deep search and let it run overnight.

DEVONagent, for example, may check 40,000 pages using more than fifty plug-ins, and then build a 200-item annotated list, sorted by relevance and saved in RTF (rich text format).

- Copernic for Windows
- DEVONagent for Mac OSX

   (See Appendix B for URLs)

Libraries and librarians

In books and journals, information has been researched, qualified, digested, filtered, and organized by writers and editors as part of the publishing process. Besides offering books and journals, some libraries enable you to search proprietary databases from your home or office, using your library card number to log in.

Librarians know their collections. Talk to librarians. You'll find they are intelligent, good listeners, very well organized, and they like to help people. They'll save you time.

Some major public libraries offer a fee-based professional research service. This encompasses the library's print collection, proprietary databases, public access databases, and the Internet. Qualified librarians do the work, and they have their full team behind them. In a major library, this can mean dozens of Masters-level librarians, all experts in their own collections, using each other's expertise. The key here is to precisely define your search.

One example is IntelliSearch at the Toronto Reference Library:

(See Appendix B for URLs)

You determine your budget and deadline. A search may progress in successive stages, with you adjusting the path as findings emerge. Results are usually faxed or e-mailed and include copies of print materials as well as electronic documents. If your local library doesn't offer this service, try a Google search "public library AND fee-based research."

Industry norms on critical issues

It's important to monitor your industry or profession for accepted norms on critical issues. This will show whether your organization is a leader, merely on solid ground, or lagging behind. You should have a short list of issues you track and assess with a SWOT analysis:

- **S**trengths
- **W**eaknesses
- **O**pportunities
- **T**hreats

On many issues your goal will be excellence as a manner of principle. But goals and performance are not the same. It's good to know the norms and how your organization ranks among peers. (See also **Narrowing the issue** in Chapter 9.)

Peer networks

Consulting with peers gives you a sounding board and a network that benefits media relations. If you are interviewed for a feature story, the journalist likely will be looking for multiple sources. When you can suggest one or two good people to talk to, you show you're connected, you build goodwill, and eventually you receive reciprocal referrals.

It's good to have a well-tended list of peer contacts, representing expert opinion on a gamut of issues. (See also **Broadening the issue** in Chapter 9.)

Walking around

There is no substitute for 'management by walking around.' Words are only a small part of communication. People can tell far more by action—by what is done and not done, by timing, by body language. People will share concerns, observations, and anecdotes *in person* that they have neither the comfort nor time to commit to paper.

When you are really listening, this comes through in face-to-face communication. People are more likely to share what really matters to them. This is essential information for effective issues management. Getting around doesn't require a large budget outlay. It does take interest, commitment, and time. Politicians know the value of this. Mixing at the community level keeps them in touch and gives them a sense for the real issues.

Focus groups, polls, surveys

More formal research is something to keep in mind at budget time:

- A major bank may hold one hundred focus groups across the country, in two runs spaced six months apart, to gauge public opinion in preparation for a major decision

- An association may buy an inexpensive question in an omnibus poll to get a broad reading on public attitudes—some online resources:
    - What is an omnibus survey?
    - ORC Macro
    - Ipsos Reid
    - GMI Poll

  (See Appendix B for URLs)

- A small nonprofit may survey its constituencies to get feedback—it's easy to do this using Web-based survey tools:
    - Survey software directory
    - "Conducting Web-based Surveys" (article)
    - SurveyMonkey (one example of a Web-based service)

  (See Appendix B for URLs)

It's important to understand the respective roles of qualitative and quantitative research:

- Qualitative research lacks statistical significance but can help identify the right questions.

- Quantitative research offers confidence in results so you can map change.

  (See Appendix B for URLs)

Informed decisions

A wealth of information is available through good reporting systems, proactive face-to-face meetings, and existing data mined from public sources. Formal research is merely a supplement. Good listening and intelligent follow-up can enable you to make informed decisions that prevent or help mitigate crises.

You may want to consider an ethics audit of your organization as part of due diligence.

Business ethics—online resources

- Ethics Resource Center
- Center for Ethical Business Cultures
- International Business Ethics Institute (IBEI)
- *Business Ethics* magazine
- Corporate Governance
- *Complete Guide to Ethics Management*

    (See Appendix B for URLs)

◆ ◆ ◆

Issues management—Questions for discussion:

1. Have you developed a list of issues that impact on your organization?
2. Do you monitor these for status and trends and do a regular SWOT analysis?
3. Can you tie issues to stakeholder groups and then segment these groups?

4. Do you have an ongoing communication program for each key stakeholder group?

5. Do you have up-to-date contact information for key stakeholders?

6. Is issues management a regular agenda item for your top-level meetings?

7. Is leadership and coordination for this area assigned to a senior manager's portfolio?

8
EMOTIONAL INTELLIGENCE

Emotional intelligence is so often ignored that it deserves a short section on its own.

Feelings are as real as facts. You learn about them by listening to and observing people.

Some perspectives

1. *People need to vent strong feelings before they are ready to listen.* When people are upset, they have a passion to be heard. "Having something to say" is a single-minded state that focuses all of one's energy and attention. Being a good listener is an effective way of being heard. This is not just a token technique. You must listen in good faith.

2. *A decision is made when it feels right.* Whether you feel right about something is a test in making a decision. No matter how logical a decision may seem, it's made firm when you feel right about it, even if this just means you are comfortable that you've exercised due diligence regarding the matter.

3. *Successful politicians know how people are feeling.* Politicians survive by staying in touch with public opinion. While they sometimes use public opinion polls, most often they just mingle with their constituencies, listen to people, hear what matters to them, and see how they perceive things.

4. *Empathy helps you speak someone's language.* To reach people you need to speak their language. You need to explain things in terms that make

sense *to them*. This is the old sales adage of speaking benefits, not just features. Perceptions and values are facts. Emotional intelligence helps you to understand someone's language and speak in meaningful terms.

5. *Stakeholders and publics are not the same.* Publics form themselves, not by logic or demographics, but according to how people feel about an issue. The process is not unlike a weather system forming. Tuning in to where people are coming from, being a good and active listener, works much like radar.

6. *Relationship skills are tied to management success.* Emotional intelligence is an important management competence, a fact confirmed in thirty years of empirical studies. The same skills apply to building relationships with the larger community. Public relations is a management function.

7. *Engagement is important for learning.* Paul Tillich, the famous philosopher and theologian, spoke of "the pedagogical error of throwing answers like stones at people who did not ask for them." Two-way dialogue involves listening as well as speaking. It engages people by stimulating the give-and-take of questions and answers. "Empathy and advice" is the mantra of adult education. This flows best from engagement and dialogue.

◆   ◆   ◆

Emotional intelligence is immediately apparent in face-to-face communication. Two different people can say the same words yet create different impressions. A large measure of a message's meaning comes through in nonverbal ways.

Nonverbal messages

Take an example from music. A student pianist and Glenn Gould both play the same Mozart sonata, but the message is clearly different. Like musical notes, words have merely the capacity for meaning. A good part of any message is carried by its setting and the way it's delivered. Body language and qualities of voice reflect the meaning for the speaker.

Transactional analysis (TA), developed by psychiatrist Eric Berne, holds that only 7 percent of a message is verbal—38 percent comes through in the way the words are said, and 55 percent is contained in facial expression.

Face-to-face meetings, where possible, give you a more informative reading of where people are coming from. This sensibility is invaluable in a crisis situation.

Debriefing

Emotional topics, common to crisis situations, are a minefield of perceived insensitivities.

It's important to know how you feel about what you're saying. It's prudent to get feedback from others as to how you're coming across. You may want to ensure that someone known for his or her social perceptiveness attends presentations, observes both the speaker and the audience, and does a debriefing afterward.

Approach

When discussing an emotional issue, it's important to acknowledge feelings.

Once that's been done, a useful precept of TA is that "I'm OK, you're OK," is an approach that works. People generally respond to respect by returning it, and a rational approach tends to engender the same.

(See Appendix B for URLs)

This suggests a version of the Golden Rule:

- Listen to others as you would have them listen to you.
- Respect their position as you would have them respect yours.
- Consider their points as you would have them consider yours.
- Deal with them as you would have them deal with you.

Doing this demonstrates what you're expecting and sets the tone for interaction.

When a discussion is becoming derailed by ideology, conflict mediators suggest that moving the conversation toward practical interests can put things back on track.

Trust

During a crisis, outsiders will be interacting with many people within your organization, not just the spokesperson. Ownership carries a tangible quality with it. When your people know their roles and are confident that a team is behind them, this comes through and encourages trust.

◆   ◆   ◆

Emotional intelligence—Questions for discussion:

1. Does your organization have a program for feedback on issues?
2. Is there a proactive approach to establishing two-way dialogue?
3. Are people encouraged to express their feelings without reproach?
4. What audiences are likely to have strong feelings on various issues?
5. What are some first steps toward initiating discussion on issues?

6. How can emotional intelligence be established as a value of your organization?

7. Are there ways to reward people who lead in developing this value?

9
FOCUS

Priorities and focus are critical in crisis management:

- Your first priority is dealing with the crisis itself and people who are affected.

- Your second priority is communicating effectively about the crisis.

Both internal and external communications are important:

- Internal communication is critical for mobilizing your own resources to serve your top two priorities.

- External communication brings more resources into play and gives the public information they need in order to take appropriate action.

The environment will be turbulent. Information will be scarce. There will be conflicting reports. A lot of people will be asking many questions.

To respond to queries, you will need two things:

- You must have something to say.

- You must have well-defined limits as to what you're prepared to say.

Having something to say

A meta message is a theme that underlies and comes through across multiple messages. It describes an impression your audiences will gather over time. If you define your meta messages and internalize them, they will drive how you communicate and what you have to say.

A message will serve you exactly as you serve it. Credibility means being worthy of your message.

Here are some meta messages you may want to consider:

1. We are aware of the situation and are responding to it.
2. This is an urgent matter for us and we are on top of it.
3. The human situation of the people involved is important to us.
4. We are competent, accountable, and responsive.
5. Other responders are working with us as a team.
6. We don't speculate or assign blame.
7. We will keep you informed as more information becomes available.

Your actual message will be specific and have concrete details—what you have to say, spoken in your own voice.

There is something to the image of a person scribbling talking points on the back of a paper napkin. This informal material gets past your internal editor and makes it easier to avoid writer's block.

Take some letter-size scrap paper and fold it in four, clean side out. In point form, jot down the top-of-mind things you want to say. Edit this down to three items.

Unfold the paper and assign one panel to each key point. Develop your points by listing top-of-mind details you want to cover under each. You should be able to do this in ten to fifteen minutes.

For each of your major points and every detail, here's what you should be prepared for:

- A concise response (15–25 words or under 10 seconds—news conference Q&A)
- A sound byte (50–80 words or 20–30 seconds—typical for TV news)
- A short discussion (150–250 words or 60–90 seconds—a topic in a radio segment)

If you can handle the quick response and the sound byte, the short discussion is just an extension of these with additional details.

You want to get right to the point, and then be prepared to flesh out your information with concise details. If you can't do this, the moment will pass before you're done—the fumble will also distract you.

Major themes drive key questions

Journalists will have questions relating to these major themes:

- What's happening
- How it's affecting people
- How it's being dealt with
- Mitigation and recovery issues
- Prevention and accountability issues
- Context to place the story in a bigger picture

It's important to be flexible and transparent. It's also important to know what your key messages are, where you want the discussion to go, and any territory you want to avoid.

If an interview is taking you somewhere you don't want to go, don't apply the brakes—steer in a new direction. You can't do this if you have nothing to say and are simply responding to questions.

Piloting a boat is a good analogy. If you don't have any forward motion—if you're "dead in the water"—you'll just bob every which way and be pulled around by the wind and current. If you have power, whether it's by engine or by sail, you can steer a course.

Hitting the brakes is somewhat like dropping anchor. It doesn't take you anywhere. And it can portray you as being defensive.

Having something to say gives you power to head somewhere. Consider your meta messages. Let them generate talking points specific to your crisis situation. Get as much information as you can. Prepare your thoughts. Internalize your key messages. Make them your own.

This preparation will give you focus and you will feel more confident speaking with the media. Remember, the better part of any message is nonverbal. It comes across in body language, tone of voice, the obvious meaning that words hold for someone. This is perhaps most apparent in radio or television coverage. But it comes through in print as well: it finds its way between the lines.

Leaders are driven people. The meaning of what you have to say will drive you.

Broadening the issue

There may be times when your organization is in crisis because of an issue that challenges an entire geographical area or industry or profession. You can point this out and broaden the issue.

If your organization is following best practice, this will come out in explaining the challenge and how organizations are attempting to deal with it. You can refer journalists to other sources who can verify this and offer their own examples. This can help build public support for changes, such as funding or legislation that might prevent or mitigate future occurrences. (See also **Peer networks** in Chapter 7.)

Narrowing the issue

Conversely, you may want to narrow the issue. An entire industry or profession can be unfairly branded because of a few organizations not following best practice. You can point out that other organizations (yours, for example) face similar challenges without incident, because you are accountable and follow best practice. This will help narrow the scope of the issue.

If an issue has been painted with a broad brush, you may be implicated without having been interviewed. A letter to the editor or producer (e-mailed or faxed, with a copy to the journalist involved) is one way to approach refocusing the issue. You should be well prepared for immediate follow-up and a media interview following receipt of your letter.

Removing your organization from implication is not the same as blaming others. This is a sensitive area and you need to be clear about what you're doing. You are not commenting on another organization; you are commenting on yours and being proactively accountable. Every organization is left to account for itself. (See also **Industry norms** in Chapter 7.)

Advocacy

A crisis is an opportunity for public discourse on public policy and the public interest. Advocacy groups and associations have a window of intense media attention in which to make their case.

Debate on a controversial issue has all the elements of a great story—dramatic tension, consequences that matter, the chance of transformation, and character change. (*Story*, by Robert McKee, offers deep insight into what makes for riveting storytelling.) This is a trial in the court of public opinion, where various groups who hold differing positions on issues advocate and make their cases.

The point here is that effective crisis communications is important for anyone who cares about an issue, not just for organizations whose reputation may be threatened. Advocacy groups play an important role in fostering debate as to whether the public interest is served.

Part 3 of this book, **Issues and Action**, can be used as a resource for both sides of the debate.

Having well-defined limits

Many people have difficulty saying no. In crisis communications, the ability to say no is important. People will ask for information you don't have. They'll speculate and ask you to agree. They'll blame someone and ask you to agree. Don't.

Very often you can get around saying no by offering an alternative:

- We'll let you know as soon as we have that information
- That would be speculation and we owe it to people to stick to the facts
- When this is over there will be time to review accountability, right now …

If you develop a clear philosophy on the issues of speculation and blame, you will find it easier to assert your limits. There's a wide variance among the media and how they pursue a story. The spectrum runs from award-winning journalism to crass sensationalism. You'll have to deal with all of it.

Know your limits and others will soon know them, too.

♦   ♦   ♦

Focus—Questions for discussion:

1. Can you think of some top-of-mind metamessages for different scenarios?

2. Have you clarified your position regarding speculation and blame?

3. How would you get your team on the same page regarding key messages?

4. How would your key messages filter down through your organization?

5. How would you coordinate your messages with those of other organizations?

6. How would you update and verify incoming information?

7. How would you provide context to the media for various scenarios?

8. Do your spokespeople have experience in stand-up, deadline-oriented situations?

9. Do you have a portable recorder kit to support speaker development?

10. Could you organize peer circles or workshops for spokespeople to get some practice?

♦   ♦   ♦

ENDNOTE

McKee, Robert. *Story: Substance, Structure, Style and the Principles of Screenwriting.* New York, New York: Harper Publishers Inc., 1997. This book offers brilliant insight into the story process and how it captures and transports an audience. Though *Story* is written about screenwriting, the

CBC recommends the book, as its principles apply regardless of medium. They also explain why crises provide such powerful story material.

10
NEWS CONFERENCES

It's critical to keep in mind two types of news briefings—internal and external. Each has a specific purpose. Do not underestimate the value of internal briefings; if neglected, you can find your ship sinking under you. (IMPORTANT: See also **Chapter 5**—Process—Breaking news)

Employee orientation

Before we talk about the mass media, it's important to remember the people within your organization. Anyone from the outside who deals with you will likely deal with some of them first, and their attitude will reflect on your organization. Your people should have a sense of urgency and purpose. They should know what's going on.

Everyone within your organization should both understand and trust the communications function. If you have an orientation program for new employees, your communications manager should be on the agenda. Employees should understand the role of public relations, how this translates into action on an everyday basis, what to do with media calls, and the role they play as sources for internal communication.

Internal sources

As a crisis unfolds, internal communication is critically important; it is the source for information public relations will verify, filter, and package for release to the public media. Your normal internal communication program is an opportunity for building trust in your public relations function. These relationships are just as important as relationships with external editors and journalists.

Internal news briefings

During a crisis, consider a brief news update for your staff at the start of their shift. This can be more or less frequent (every few hours or every few days), according to the situation and your news. You can update your people with the latest information, thank them for their dedication and hard work, answer their questions. This keeps them in the loop, helps them feel they're in on things.

Whenever you issue a public news release, immediately post the same information internally and notify supervisors to announce the posting. Employees shouldn't need to learn about their own organization from the radio.

Where there's strong interest but little information, speculation and rumor will fill the void. Better to spend your time building trust than repairing damage. You have an opportunity to establish management as a credible and approachable news source. This is something you should not ignore.

Mass media

A media deadline is somewhat like a train schedule. If you are late, the train will leave without you. The presses will roll and broadcasts will go on air with or without your timely participation.

When there's strong media interest in a story, you can be overwhelmed by interview requests. A news conference enables you to answer questions from a room full of journalists, all at one time.

There are three things you'll want to have as you go into a news conference:

- Deep background on the issue—gained from briefings, notes, files

- Close contact with the scene, immediate reports from people directly involved
- Consensus from your crisis team on the situation and talking points

External news conference

Your first choice for getting your story to the media should be a news release. This is the most efficient way to get information to newsrooms. Journalists are busy people and they work in a highly competitive market. Traveling to a news conference takes valuable time that must be offset by special value provided by the conference.

A news conference works like a group interview and may be justified when these conditions are met:

- Your story is highly newsworthy—you are swamped with media attention
- A conference would enable you to make available special resources like key players and experts
- A conference is clearly the most effective and efficient way of informing the media
- An open conference may be the only way of handling the volume of interview requests

Location

You will need a place to hold your news conference:

- Choose a venue that's convenient to newsrooms. This will likely be "downtown." Your offices, a hotel conference room, your city's press club, or a borrowed room at city hall are all possibilities. Your mayor's office may have suggestions.
- Unless a background sound is part of the story, avoid locations where aircraft, trains, and streetcars roar, rumble, and screech. A loud building

ventilation system that cycles on and off can also pose a problem—ask if anything can be done about this.

- Contact one or two audiovisual services to see what's involved in setting up a group feed, which runs an audio feed from the podium to a distribution box at the back of the room. Broadcast journalists plug their recorders into the common feed and you don't have a dozen microphones in your face.

- Someone should know which power outlets are on which circuit breakers and how to quickly restore power. This person should direct TV crews to where they can connect their lights. Two 1000-watt TV lights on the same circuit will cause an overload and delay your start.

- A close-by staging room can be very helpful in providing your people some privacy to prepare for the conference. You can also use the corners of this room for one-on-one interviews following the conference.

- A dry mouth is something that happens in the real world when people face an audience. Water is essential for the podium. Lip balm may be appreciated. Pocket packets of facial tissue can be handy, if only to clean eyeglasses. Your spokespeople should feel well tended.

- If the spokesperson is using a printed text, keep these rules in mind: Use only the top half of the page to make it easier for the speaker to face the microphone and the audience. Keep the pages loose, so the speaker can slide them to the side as each is finished. Large page numbers at all four corners (use headers and footers) make it easy to shuffle pages. Use 14-point type and 1.5 line spacing. This will give you 160-180 words or about one minute of speech per page. Normal capitalization (not all caps) is easiest to read (the ascenders and descenders of lower case letters give the eye cues to recognize words).

- Record an MP3 file or an extra tape of the conference for latecomers who were caught in traffic. Immediately following the conference, they can gather in a corner and fill in their notes from the first few minutes of the recording. (This recording will be handy for your records as well.)

- You'll need long tables in the hallway for registration. Mark people off against those you were expecting. E-mail the URL for your media kit to those who couldn't attend (but not until after the event). Note who's late and mention your audiotape to them.

Timing

Timing that works for both you and the media is important:

- Schedule your conference so it will give journalists time to meet their next deadline. Unless they came with a satellite truck, TV crews will have to travel back to their stations. Print journalists need to write their stories and file their copy, which then needs to pass through editing and production to make the next edition. Phone the newsrooms at your major daily paper and a few TV stations. Ask them which times work best.

- Unless there are other considerations, a mid-morning or early afternoon news conference avoids rush hour traffic and should be convenient for most deadlines.

- Try to establish a pattern for news conferences and releases. Let people know when to expect your next update. You can override your forecast if there are unexpected developments.

Media advisories

A media advisory works like a news release, except it's an invitation to attend an event. An advisory is a very brief half-page release and should contain:

- Greeting: Attention News/Assignment Editors

- Title: Media Advisory—followed by your headline

- A short paragraph on the purpose of the news conference—finish by saying the media are invited

- Date

- Time
- Place
- A note on any photo opportunities—important for television and newspapers
- Contact information

Media handouts

The news conference kit should include:

- News release
- A "big picture" background piece on your organization
- A "window" background piece on any program/division involved
- Fact sheets (backgrounders) on the key issue, if relevant
- Reproduction-quality copies of any graphics or illustrations (or a URL for downloading these from your Web site)
- Concise biographic sketches on your spokespeople and/or panelists
- A glossary of acronyms and technical terms, if this is relevant
- Contact information

You can make the kit available for download from your Web site—immediately following the news conference. Don't make the information available ahead of time, as that would give those who don't attend a head start.

Graphics, photos, or illustrations that are part of a media kit should be reproduction quality—at least 300 dpi resolution and larger in size than their final use requires. In most cases, this means a 1,500 x 2,000 pixel JPEG file. Post thumbnails with links to full-size files. It's a good idea to not link to this page from your public area. Instead, include a direct URL to the download page in your media releases.

MP3 files

You may want to be able to offer a sound file to journalists who've arrived late at a news conference or to post a sound byte to your Web site.

If you are getting a digital recorder, here are some things to look for:

- MP3 recording capability
- Removable media, e.g. Compact Flash or SD card
- USB 2.0 connectivity
- Field replaceable batteries, preferably AA
- Robust software, few problems reported
- Physical durability

As of this writing, the Edirol R-09 and the Marantz PMD660 are two professional MP3 recorders that meet these criteria. Each costs about ten times as much as a microcassette recorder and consumes batteries at about five times the rate. Manufactures' marketing claims are fine, but look at user forums to see what experience people are having. User reviews at Amazon.com and the radio freelance site Transom.org are credible sources for equipment reviews.

You can fit six hours of CD quality sound (320 kbps) or thirty hours of interview notes (64 kbps) on a 1 GB memory card. When you stop and restart an MP3 recorder, it begins recording to a new, numbered MP3 file. You'll find yourself mindful of battery life to record for the time you require. A small memory card reader is convenient for transferring files to your computer.

Intuitive and inexpensive shareware—MP3 Trimmer for Mac OSX and MP3 EasySplitter for Windows—allows you to review MP3 files and extract sound bytes without losing quality. If you take point-form notes during an interview, noting the running time, you can go directly to that point in an MP3 file to listen or edit.

(See Appendix B for URLs)

Panel and presentation

You may want to gather a few people to form a presentation or resource panel. Each panel member can take a few minutes to present, or the key spokesperson can present and then direct questions to panel members according to their specific expertise. In either case, everyone should know the key talking points and messages that are important to get across.

Telephoto lenses and shotgun (long-range) microphones have a long reach. Don't place notes in plain view that you wouldn't want photographed and published. And you must assume that anything you say, even as a private comment to the person sitting next to you, can end up "on the record."

Try to provide graphs, maps, and pictures if they are relevant to your message. Consider whether you need a release for identifiable people in photographs. Talking heads are fine for radio, but television and newspapers will welcome visuals.

If you use visuals, you'll need an easel and a pointer. The easel should be within easy reach of the podium so the speaker doesn't have to leave the microphones to point to something. Someone should be available to change the visuals. You may want to use a cover sheet to avoid distraction until the items are needed. Try all this ahead of time.

Duration

If one person is speaking, ten minutes or so of presentation followed by twenty minutes of questions from the floor works well. If several panelists are to speak, try to limit their initial remarks to no more than five minutes each. The spokesperson may want to read a short prepared statement. You can refer reporters to material that has been supplied to them in their kits. Interaction is the highlight of a news conference. You want to quickly get to the questions.

Ten minutes of remarks is roughly 1,600–1,800 words. Sketch your talk: your core message, three to five main points, details to cover under each point. Be articulate in less than one hundred words (sixty or fewer is best) on every point you want to make. Your presentation will build very quickly.

Try to speak to your points without reading. Use point-form notes to keep yourself on track.

If you have the time, try a practice run. Tape yourself and take note of the time spent on each topic to help you decide where to trim or add material. Cut back what sounds like fluff. Add important details you notice are missing.

During the conference, have someone at the back of the room to keep track of time and signal speakers so they know their time status. It's a good idea to have another person record the conference and jot down all questions asked.

Prepared statements

If you have a formal statement, read it at the media conference. The journalists will follow along in a printed copy and note any significant departure in content. Before the conference, read your statement aloud for practice, and then rewrite it so it sounds like your spoken word. A formal statement shouldn't be long—100 to 200 words are usually plenty.

Remember that the broadcast media will carry your statement, and that the text is only a fraction of the message carried by your voice. Focus, presence, sensitivity, energy, and confidence—all these qualities come through between the lines. Internalize your message. It has to come from *you*. You should not sound like a computer reading text. You need to sound like a caring, competent person who's on top of the situation and knows what to do.

Questions

When you've finished your presentation, let the audience know you are ready to take questions. People should identify themselves and who they represent when you recognize them. Make your answers direct and brief. Keep your key messages in mind and incorporate them wherever you have an opportunity. If you don't have the information someone is asking for, don't speculate. Tell them you don't have that information but will supply it when you do. If someone asks a negative question, don't repeat it when answering.

When the scheduled end of the session is approaching, let the audience know you have time for one or two more questions. When the time is up, end the conference. Everyone has work to do.

Have someone prepare a list of the questions asked and answers given. After the conference, work your way through the text and improve your answers. Make them concise, accurate, to the point. About sixty words (there are exceptions) is the outside limit for an answer. A general rule is: the shorter, the better.

If you have the time, *Asking Questions: The Art of the Media Interview* is an outstanding book by Paul McLaughlin. It was written for journalists but is equally useful to news sources. McLaughlin explains in detail the process of the media interview and covers the differences between print and broadcast.

Breakout interviews

You and your panel members can make yourselves available for interviews immediately following the conference. If you have a staging area, use it for interviews with print and radio journalists. Television will often want to do on-the-spot interviews just outside the conference room. You might want to consider which areas would offer a good visual backdrop. Schedule these interviews in advance, and stick to your time schedule. Journalists need to talk with you, then get back to file their stories.

Microcassette recorders

Just about everyone uses digital technology for listening to music. But until you see journalists generally using digital recorders, tape recorders should be considered the most reliable and practical.

Get a good but inexpensive microcassette recorder and a supply of tapes. Tape your presentations and interviews. Reviewing the recording can help you see which points you've made and how you can improve. Recordings also serve as backup in case you are misquoted or your words are taken out of context. Microcassette recorders have two speeds. If you use the slower one, you can get sixty minutes per side on a standard tape. Mark the label with a simple start-time code—YYMMDD–HH:MM. This gives you the exact date and time of a recording that is easy to match with your list of interviews.

Most microcassette recorders use one of two mechanisms, and the counter differs on them. A chart in Appendix A shows counter readings versus minutes into a tape for Sony and Olympus. Your microcassette recorder will likely agree with one or the other of them. If you or someone else takes notes on topics and time, it's easy to translate this into minutes from the start of an interview and quickly find that spot on a tape.

(See Appendix A)

A good microcassette recorder kit should include these items:

- A Sony, Panasonic, or Olympus microcassette recorder that uses AA batteries

- A dozen Sony microcassette tapes (and some Avery 02209 labels)

- A telephone pickup (see Radio Shack or Circuit City)

- 1/8" or 3.5mm dual headphone adapter (or mini headphone splitter or Y adapter)

- 1/8" or 3.5mm extension cord with male-to-male connectors

- Spare batteries (always have a set of alkaline batteries as a backup for rechargeables)

Also, a Crown Sound Grabber II microphone can be a compelling tool for capturing everything, including panel speakers and questions from the floor. Crown holds the patent for PZM (pressure zone microphone) technology, which picks up clear, bright sound with remarkable detail—from an entire room. The Sound Grabber is inexpensive for the performance it delivers, and it plugs directly into a microcassette recorder or computer. When it's not being used in news conferences, it can earn its keep as an excellent mike for VoIP (voice over Internet protocol) conferencing or recording group meetings.

(See Appendix B for URLs)

Journalists are heavy users of microcassette recorders. You'll see most of them using an inexpensive Sony model that takes AA batteries. AA batteries cost half as much as AAAs and have three times the ampere/hour (A/h) rating. The little Sony recorders have withstood the test of hard reportage.

If you're using rechargeable batteries, the MAHA brand is consistently recognized as a leader. Their chargers revive and condition NiMH (nickel-metal hydride) batteries so they perform at full capacity.

Regular NiMH batteries lose their charge at a rate of 1 percent per day. To keep within 70 percent of capacity, you would need to charge them every month. A new generation of NiMH batteries greatly reduces self-discharge. Sanyo claims its Eneloop brand will keep 85 percent of a full charge for twelve months. Panasonic says its R2 brand will keep 80 percent for six months. Both batteries come ready to use and can be recharged a thousand times.

(See Appendix B for URLs)

A dozen microcassette tapes can hold twenty-four hours of interviews. Avery 02209 (S828N) labels, cut lengthwise, are perfect for labeling the tapes. (A transcriber with high-speed erase is convenient for recycling tapes.)

The telephone pickup enables you to record media phone interviews. (The Olympus Mini Tele-Recording Device works with the headphone jack of some cell phones.) Recording phone interviews with the consent of at least one party to the call is usually legal in the United States and Canada. Some states require two-party notification. If this is your case, just mention to the journalist at the start of the interview that you always keep a tape.

CallCorder.com offers a synopsis of the regulations for the United States and Canada.

(See Appendix B for URLs)

The dual headphone adapter and male-to-male extension cord allow you to record a radio broadcast while listening to it. Just run the extension cord to the MIC jack on the recorder. On many recorders, you'll have to adjust the recording level by using the volume control. You'll need to play with this for the best setting, so make sure you do a "dry run" before relying on the technology.

When you start a fresh tape, always push the counter reset button to "000." When you review a tape, take point-form notes that begin with the counter reading.

A microcassette recorder is a great learning tool. It lets you sit in the audience after you've spoken.

◆ ◆ ◆

News conferences—Questions for discussion:

1. Is media coordination an established function within your organization?

2. Can you identify some needs for both internal and external news updates?

3. Where would you hold a news conference? Who would handle the venue logistics?

4. Who would handle logistics for hard copy media kits and downloadable Internet files?

5. How would you handle news updates to your Web site? Is there a fallback plan?

6. Do your managers have public speaking experience in a deadline-oriented environment?

7. Have you identified conference panel resource people for some scenarios?

8. Do you have a tested system for monitoring media coverage?

9. How would you gather and manage frequently asked questions?

10. How would you track, prioritize, and handle media requests?

11
MEDIA ACCURACY

In a crisis, you can't work from a recipe book. You must be able to think on your feet. Observing some simple rules and following up if you find mistakes will help ensure accurate media coverage.

This brings us to the ubiquitous *Dos and Don'ts* list.

Nine Crisis Dos

1. Do establish media coordination as a trusted process *before* a crisis occurs.

2. Do be responsive to media queries and try to meet deadlines.

3. Do establish a pattern of releasing information; synchronize this with broadcast deadlines.

4. Do release the same information to all media; maintain an even playing field.

5. Do focus on progress and achievements while acknowledging suffering and loss.

6. Do monitor media coverage and promptly but courteously call attention to inaccuracies.

7. Do be prepared to provide context for the crisis situation.

8. Do be prepared to refer journalists to other useful sources.

9. Do thank everyone who is helping—be as specific as possible.

Nine Crisis Don'ts

1. Don't wait hours to respond.

2. Don't go into an interview unprepared.

3. Don't release personal information.

4. Don't blame; don't let anyone entice you to blame.

5. Don't speculate; don't let anyone entice you to speculate.

6. Don't mislead *or* cover up facts *or* lie *or* be disingenuous.

7. Don't let incorrect or misleading statements pass without comment.

8. Don't make "off the record" statements (promises of confidentiality are often forgotten).

9. Don't repeat negative or inflammatory words—they might end up as your quote.

Monitoring the media

Large organizations can afford media monitoring services, but they can be expensive for smaller organizations, and service often takes a few days to set up. If you leave monitoring until a crisis occurs, you'll likely miss the early coverage.

When you begin your search for a monitoring service, start with suppliers of media lists. You'll also find resources on the Web.

(See Appendix B for URLs)

If you don't subscribe to a monitoring service, the following resources can help you track coverage.

Online news sources:

- Journalism.net—Canada and United States pages
- American Journalism Review—tracks multiple sources
- NewsLink—U.S. and Canadian media
- Google News—search and browse 4,500 news sources
- Yahoo News—keyword search across multiple media channels
- Disaster Center—newspaper, television, radio links—United States

Broadcast URLs—streaming media feeds:

- Radio and Television Stations on the Web—worldwide
- Radio Locator—United States and Canada

Print news sources:

- Newspaper Association of America—rich resources
- Canadian Newspaper Association—rich and varied resources
- World Press Review—perspectives from around the globe

   (See Appendix B for URLs)

Media tracking is unlikely to be effective unless you have a plan. At the very least, assign the tracking of newspapers and radio and television stations to specific people. Everyone involved should have a separate agenda and a good supply of reliable VHS tapes or DVD recordable media.

It's important to have a labeling and indexing protocol. It's good to take point-form notes while a newscast is playing, showing the time the recording started and a clock time beside each point. This provides a quick index for finding and checking content. For most flexibility, organize notes in a Microsoft Excel spreadsheet with fields for station, date, and time. You can then review coverage by media type, station (cross-referenced with station demographics), time, keyword, and other appropriate criteria you can add.

If resources are limited, it makes sense to monitor your local media, the national media, and major media in four dominant population areas. Half the population of North America lives within a two-hour drive of New York, Chicago, Los Angeles, or Toronto.

Panasonic VHS and DVD recorders are consistently rated as the most reliable. Sony has the best reputation for reliable recordable media—VHS videotapes and blank CD and DVD discs. An antenna splitter (available at Radio Shack or Circuit City) will enable you to use three recorders to simultaneously record three major networks. Just turn the recorders on one at a time manually to program them with the remote.

DVDs have a number of advantages over VHS videotapes:

- DVDs provide date-and-time-stamped thumbnails for quick and easy navigation.

- Even a basic Panasonic DVD recorder can handle ninety-nine separate items per disc.

- It's easy to pause and fast forward or skip back and forth between items.

- The storage media is very compact and inexpensive—six to eight hours per disc.

- DVDs can be played on computers (the disc needs to be "finished" or "closed" first).

One disadvantage of DVDs is that if you get a defective disc (known as a "coaster") it won't record. If you buy only DVD discs that have been proven to be the most reliable (for example, Sony), it's unlikely you'll have any trouble.

Whether you're using videotape or DVD, buy an inexpensive UPS (uninterruptible power supply) unit and use it to power your recorders. Power irregularities are a fact of emergencies, and even a brief one will require you to reset all your settings.

Google Alerts

Google Alerts can work as a simple, worldwide media monitoring service.

1. Set up a Google Alerts account and ask it to notify you when a keyword or phrase appears the 4,500 news sources it monitors.

2. Put Google Alerts in your address book so your system will recognize it as a known e-mail address.
   Name: Google Alerts
   E-mail address: googlealerts-noreply@google.com

3. Set your SPAM filter's whitelist to allow e-mails with "Google Alert" in the subject line.

Alerts will have the search term in the subject header, e.g. "Google Alert—your_search_phrase". Some e-mail clients allow you to filter mail by subject into a designated mailbox or message folder. It's worthwhile to set this up.

To access a story cited in an alert, you'll need to view your e-mail as HTML and click on the story's link. Windows users can identify recurring phrases in media coverage by using Textanz—inexpensive shareware that analyzes text and produces a list of words and phrases with their frequency of use. Textanz also allows you to export its reports to Excel. If you have an Intel-based Mac, you can run Windows applications without rebooting by using Parallels Desktop.

(See Appendix B for URLs)

DNA13

DNA13 offers very powerful (and priced accordingly) real-time issues and media monitoring software that's used by some of the world's largest corporations. If you have large revenues, the benefits of DNA13 may make economic sense.

(See Appendix B for URLs)

Facilitating media accuracy

The BBC makes some excellent points on how to develop a relationship with the media:

(See Appendix B for URLs)

Stylebooks and codes of ethics can give you a sense for media standards:

- *Associated Press Stylebook* and *Broadcast News Handbook*
- *The Canadian Press Stylebook* and *Broadcast News Talk*
- Media Codes of Ethics—Canada, United States, and other countries
- Ethics in Public Broadcasting (United States)

(See Appendix B for URLs)

Journalists and editors can make honest mistakes. Assume they are acting in good faith.

Some online resources:

- Regret the error
- American Journalism Review: Errors on air
- Columbia Journalism Review: Handling corrections
- American Society of Newspaper Editors: Credibility

  (See Appendix B for URLs)

Research the ombudsman function (search "ombud," a gender equity issue), bureau of accuracy, and any press council or broadcast licensing authority associated with media you're likely to deal with. You'll then be ready to call their attention to any errors.

Point out errors immediately, ideally within the same news cycle. E-mail or fax your comments. Send your message to the journalist involved and copy the editor. Follow up with a courteous phone call. Listen as well as talk. Make yourself available. Ask for a correction.

Every person on your senior management team has assumptions regarding communication. To work together effectively, share views and see if you can work toward a consensus.

◆ ◆ ◆

Media Accuracy—Questions for discussion:

1. Which crisis dos are you ready for? Which would require preparation?
2. Which crisis don'ts are you ready for? Do any require preparation?
3. Do you have a process for sorting out the necessary tasks? Do you have buy-in?
4. Do you have a plan for monitoring the media during a crisis?

5. Does your organization have a formal statement outlining its public relations function?

6. Is public relations seen as a senior management function within your organization?

7. Would your organization assign a manager full time to deal with communication in a crisis?

◆ ◆ ◆

ENDNOTE

Nine Crisis Do's:
1. Do establish media coordination as a trusted process before a crisis occurs
2. Do be responsive to media queries and try to meet deadlines
3. Do establish a pattern of releasing information; synchronize this with broadcast deadlines
4. Do release the same information to all media; maintain an even playing field
5. Do focus on progress and achievements while acknowledging suffering and loss
6. Do monitor media coverage and promptly but courteously call attention to inaccuracies
7. Do be prepared to provide context for the crisis situation
8. Do be prepared to refer journalists to other useful sources
9. Do thank everyone who is helping — be as specific as possible

Nine Crisis Don'ts:
1. Don't wait hours to respond
2. Don't go into an interview without preparation
3. Don't release personal information
4. Don't blame; don't let anyone entice you to blame
5. Don't speculate; don't let anyone entice you to speculate
6. Don't mislead or cover up facts or lie or be disingenuous
7. Don't let incorrect or misleading statements pass without comment
8. Don't make "off the record" statements (promises are often forgotten)
9. Don't repeat negative or inflammatory words (they might end up as your quote)

from
Crisis Communications: A primer for teams
available at www.topstory.ca

(A PDF file for printing wallet-size cards is available at www.topstory.ca)

12
TESTING

Sports teams have practice sessions so that team members can learn how to play together.

There are at least six reasons to test your crisis communications plan:

- People learn by doing.
- You can develop ownership for the plan and its process.
- Some problems will only become apparent in a practice session.
- You can try successive versions of your plan and see what works best.
- You will develop realistic benchmarks and expectations.
- Teams that practice together work better under stress.

There are three ways to test your plan: seminar, desktop, and simulation exercises. In simple form, these involve:

- *Seminar/walk-through*—you meet to share what you have, get a response from people, incorporate new information, and manage expectations. This is useful for internal coordination and for liaising with outside organizations.
- *Desktop mini-drill*—you apply a scenario against your plan to identify issues and gaps. Participants must be able to accept the scenario as viable; this process is part of the drill.

- *Simulation/functional exercise*—you take the kernel of a plan and then use method acting to develop it. Players begin with a crisis scenario and roles and goals. They then play this out, finding barriers and solutions and documenting the process as they go.

Implementing your plan shows you what works and what doesn't. Start simple and play with everything until your plan works—then try some different scenarios.

Tracking timelines is important. To record events, it's important to synchronize time and use action journals.

Your crisis team should account for every request and response—the request itself, any waiting periods, and time spent on every related task. Software can make it much easier to collect and document this.

Time-tracking software

For Mac, Unix, and Windows users, TimeBox is a simple, inexpensive, Java-based time tracking application. It enables users to attach notes to time records, has search and merge features, can work across a network. For larger organizations, TimeBox integrates with ClokBox and TimeSuite.

(See Appendix B for URLs)

For Palm users, AllTime is an excellent and intuitive solution that exports to Microsoft Excel.

(See Appendix B for URLs)

If you're not using dedicated time tracking software, a standard Microsoft Excel spreadsheet can greatly simplify the consolidation of information.

Every step should have two time figures—best and average—and estimates for dealing with foreseeable contingencies. There should be work-arounds for steps that might take too long. This fosters ownership for results.

Encourage your staff to jot down observations, with the exact time, as soon as they occur. Thoughts have a habit of vaporizing, especially in a fast-paced, turbulent situation.

Murphy's Law

When you're ready, ask a third party to give you an arms length scenario, something you're not prepared for. Ask for unexpected barriers (Murphy's Law manifestations) that appear bit by bit as the situation develops.

Escalating system-wide breakdowns are a common feature of disasters. A power outage can lead to an immediate fuel shortage and, three or four days later, a water shortage. Mobile networks go down when fuel reserves for transmission tower generators are depleted. Dominoes fall, hitting other dominoes, expanding the number and scope of failures.

Initiate unscheduled crisis drills. A real crisis doesn't phone ahead to book an appointment.

Exercise the first hour of your plan, then the first four hours. See what works, which parts stumble, whether anything just doesn't happen. Everyone should use an audit form to track incoming and outgoing messages, itemizing action taken and response times. (See the earlier section on time tracking software.) The next day, ask everyone to file debriefing notes—what went well, any frustrations, how they felt about the exercise.

Give everything a run-through and see what works. Dig in and fiddle. Number and date your action plan versions.

Sometimes you'll backtrack because an earlier version worked better.

Establishing a critical path and raising the bar

In the end you want a critical path with roles, tasks, and timelines. All members of your organization should know what position they play and what is expected of them. Contingencies should be anticipated. There should be buy-in and a strong sense that only the team—not individuals—can fail.

Easy-does-it is the way to start. Your team needs to taste success. Once you've been through this a few times, you can challenge them to raise the bar.

To raise the bar, each unit can brainstorm about obstacles and contingencies—what could go wrong and how a situation could escalate. Write potential problems down on index cards and shuffle the deck. During your next exercise, periodically roll dice or flip a coin to decide whether you draw a card.

Team tasks

Every department or performance unit should have four items:

- A concise statement of their role in a crisis situation

- An annotated point-form list of items for which they are accountable

- Test scores on the time it takes to accomplish various tasks—best and average times are useful for managing expectations

- Notes on any likely contingencies and barriers, the typical time it takes to deal with each, and available work-arounds

Testing should have (and be seen to have) full support from the very top. Your senior managers should take the lead. Buy-in should be clear right through your organization.

Testing will give your people a real sense for the positions they play and what their team can do. They'll learn what they can expect from one another. You'll learn what you can expect from them.

You'll have a better sense for how long it takes to reach the right person, to get necessary information, to accomplish something.

Besides testing your plan, you'll be developing trust and confidence in your response team.

◆ ◆ ◆

Testing—Questions for discussion:

1. Would testing work for your organization?
2. Why would testing work for your organization—or why not?
3. What would you want to have in place before trying a test?
4. Who would be the best person to coordinate testing?
5. How would you go about planning the process?
6. Would you have support from your top leadership?
7. What budget would cover staff time required for testing?

PART 4
RESILIENCE AND CONTINUITY

A basic concept in engineering is redundant, dissimilar systems. This section can apply to both small organizations and those with sophisticated business continuity systems. Making provisions for disruptions can allow your organization to function when, suddenly, everything is different.

(IMPORTANT: See also **Chapter 3**—Resources)

13
EMERGENCY PROVISIONS

In any crisis, you will need all hands at their stations:

- The crisis response itself demands full resources.

- You need accurate and accessible sources of information for news releases.

- Communication requires support—there's a lot of detail.

- Your media relations function should be available around the clock.

Family, school, and work are basic social institutions. They are where we spend our days. Whether or not you're directly involved in a crisis, disaster response and recovery requires an entire community. "Entire" includes your organization. You owe it to your employees, their families, and your community to be prepared.

To deal with any crisis, your senior management team needs to be at work and able to function effectively. Crises are not limited to regular business hours. People will be thinking about their families.

If this is a community disaster, resources such as electricity and gasoline may not be available. If contingencies are not planned for, they become barriers. Anticipating needs and providing for them can prevent or mitigate problems.

Services are available that can notify forty thousand people in twenty minutes, based on proximity to a specific location as determined by postal code—or provide mobile offices for a thousand employees, complete with computer workstations and satellite telephone and Internet access. You'll find resources like these in the *Disaster Resource Guide* and the *Edwards Disaster Recovery Directory*, listed in the Appendix.

In comparison, the provisions covered in this chapter are low level. They are inexpensive, have universal application, and many can be implemented through personal initiative. The more these provisions pervade your organization, the more resilient it will be.

The place to start is with your senior management team. If your leaders believe in providing for the unexpected, if they have personally followed through on this commitment, preparedness will take hold as an important value within your organization.

OFFICE

Ensuring that an office can function effectively in a crisis or a disaster environment requires the outlook of a business continuity manager. You need to anticipate and provide for extraordinary conditions.

Stress breaks

Staff under stress can function effectively for only so long. Our bodies produce adrenalin in response to stress—a natural chemical that helps boost our performance, but its push doesn't last. People need time and space to recuperate.

There's the story of a senior manager with twenty years of experience who worked nonstop for seventy-two hours during a crisis situation. After twelve hours without a break, his effectiveness began to suffer. As time went on he began to make poor decisions. He was fired immediately after the crisis. The situation and how it was handled seriously affected team morale and became narrated history within the organization and throughout its peer network.

Even during a crisis, breaks are essential: a twenty-minute break every two hours and a one-hour meal break every four hours. If the situation is critical, you may be able to cut these breaks in half. People need to get up and move around, get some fresh air, take a short nap. The change of pace makes them more effective and helps to keep their effort sustainable.

With no breaks, you'll have mistakes and burned-out staff to deal with on top of the crisis. And if senior managers are involved, the entire team is impacted by their effectiveness.

Medication

People on medication don't typically carry a three-day supply with them. Running out of pills or not getting an injection can create a medical crisis for many people within a matter of hours.

You can ask your senior managers and anyone on a crisis response team to keep a week's supply of any medications they take locked in their desk. They'll need to label the medication and jot down expiration dates.

First aid

In many jurisdictions, legislation requires employers to make first aid provisions for their employees. This usually includes:

- A current edition of an approved first aid manual

- An approved kit of first aid supplies

- An employee on duty who has been certified in an approved first aid course

This is a very good idea, whether or not the law requires it.

Depending on whether you are in the United States, Canada, or the United Kingdom, one of these first aid manuals will be recognized:

- *American Red Cross First Aid: Responding to Emergencies*
- *First aid manual: The authorized manual of St. John Ambulance, St. Andrew's Ambulance Association, and The British Red Cross Society*

Both books are available through amazon.com. Best practice can change from one year to the next, so check to ensure you have the latest edition.

A manual may be useful, but training is critically important. If someone has gone into shock, just ten minutes can stand between life and death. You'll want a rescuer who has been tested on the manual and has hands-on practice in applying this knowledge.

Employee assistance

Any crisis or disaster is a time of high stress and potential trauma or loss.

If you have an employee assistance program (EAP), members of its staff should be involved in the crisis planning process. During a high stress situation, EAP counselors should be visible and available. Your employees will be working long hours under pressure and may be worried about loved ones. There may be trauma or loss. Your managers and supervisors should know what help is available. They should understand how to identify and refer someone who may need help.

It's also important to know what mental health and trauma counseling services are available in your community. Most cities have a community information center that offers comprehensive information on social services. These often have a twenty-four-hour hotline and publish a directory. But a phone number and a book are not enough. Your team should invite a social worker or a clinical psychologist to one of its meetings to discuss what people may experience in a crisis situation and how you can ensure they get quick and appropriate help.

Surviving and Thriving: Living Through a Traumatic Experience is an audio book that is available on CD or for download on the Web.

   (See Appendix B for URLs)

Produced by Dr. Mark Lerner, a clinical psychologist, this audio book features three practical sessions. You can listen to a free five-minute excerpt on the Web. This will give you some sense for the personal issues your people may be dealing with.

Employees need to understand the role they play in supporting each other. They need to understand when it is appropriate to suggest intervention of the EAP. It might be good to discuss mutual support and the role and importance of your EAP in unit level meetings.

If you don't have an EAP program, find someone you can go to for help. Meet with them and discuss their role and how your employees can be made aware of the help that is available. Most EAP programs are "broad brush" and include all social issues.

These URLs offer a good overview of EAP programs:

- Office of Personnel Management—Employee Assistance Programs
- National Institute for Mental Health
- Employee Assistance Professionals Association
- Family Services—Employee Assistance Programs

   (See Appendix B for URLs)

Eaposters.com gives you a sense for how an EAP program can be promoted through posters in the workplace.

(See Appendix B for URLs)

Support from the top on down, orientation sessions, frequent mention in your print communications—all are important for an EAP to work.

There are online resources specific to trauma and disaster mental health:

- National Association of School Psychologists
- CDC—Disaster mental health
- National Mental Health Information Center—Disaster mental health recovery
- American Academy of Experts in Traumatic Stress
- Red Cross > Masters of Disaster > Facing Fear (lesson plan for schools)

(See Appendix B for URLs)

Collective agreements

If there are unions in your organization, it's important to involve them in the planning process. Your full team needs to feel ownership for the crisis response. In extraordinary circumstances, working hours and conditions and job responsibilities will likely differ from what's normal. Even if the union has been directly involved in planning, timely two-way communication with union leaders is not to be neglected during a crisis.

Commuting issues

It may be impractical or even impossible for staff to commute. A gas shortage, a pandemic, or a 24/7 effort are just some situations where the ability to work from home can be important. Telework capability can reduce office space requirements and make your operation far more robust in the event of an emergency.

Telework/telecommuting information sources:

- The Telework Coalition (offers a benchmarking study)
- Canadian Telework Association (includes the U.S. telework scene)
- Office of Personnel Management—Interagency telework site
- Flexibility—Resources for flexible work (a UK site)

(See Appendix B for URLs)

Teleconferencing through VoIP (voice over Internet protocol) is an inexpensive way to hold meetings. Gizmo Project is a free, cross-platform, award-winning VoIP application. There's no charge to participate in a call via computer. A small charge applies if a landline or cell phone is used. Gizmo technical support will tell you how one conference call connected twenty-eight callers in eleven countries (if more than ten callers are involved, participants need to mute their microphones when they're not talking).

(See Appendix B for URLs)

An inexpensive Internet-based fax service can give your staff, wherever they may be, access to incoming faxes and hard copies of documents located at the office.

Internet-based fax services:

- eFax—e-mail fax services
- MyFax—online fax to e-mail service
- TrustFax—online fax revolution (includes fax broadcasting)

- Instant InfoSystems—RightFax fax server solution

  (See Appendix B for URLs)

Some large organizations have an arrangement with a nearby hotel or motel to provide a block of rooms on short notice during an emergency.

If you have staff whose homes are within a twenty-minute walk of the office, you can explore the possibility of their providing temporary accommodation for a few people during an emergency.

If you need to sleep at the office, self-inflating foam mats, available at backpacking stores, are remarkably comfortable if you let them inflate fully then let out a little air:

(See Appendix B for URLs)

Power outages

In the simplest of emergency scenarios there is often no electrical power—sometimes for days.

This immediately disables:

- Credit cards, bank machines, and cash registers
- Gas pumps and traffic signals
- Refrigerators and freezers
- Heating and air-conditioning

Telephones work until emergency generators run out of fuel. Traffic becomes a problem due to stalled cars and the lack of stoplights.

In an era of cordless phones, everyone should have at least one conventional hardwired telephone at home. These will continue to work on telephone line power when the main power is out.

Some large organizations have diesel generators. A fuel cell generator (by APC) that runs on hydrogen and puts out 10-30kw of emergency power is now available. Its only by-products are water and heat.

(See Appendix B for URLs)

However, backup generators eventually run out of fuel. It's good for your senior managers to have a source of backup power at home, especially since they may be stuck working there for a while.

An inexpensive 150–200-watt power inverter can charge a cell phone, a PDA, a laptop computer, or run a battery charger to replenish rechargeable AA batteries. This size (150–200 watts) is typically designed for continuous use and often includes both an accessory plug and battery clamps. An inexpensive shaving kit, available at a dollar store, makes a great storage bag.

Most inverters shut off at 11 volts to avoid completely draining the battery. Some show a digital readout of the battery voltage, the current being used, and trouble-shooting codes.

Auto supply stores are one source. You can also do an Internet search for 12-volt power inverter.

Xantrex makes a full line of inverters, including two XPower Pocket models that include both 115 VAC and USB outlets.

(See Appendix B for URLs)

If you need to recharge a device in your car, it goes without saying that you must never run a vehicle in an enclosed space. If the key must be in the ignition for the vehicle's accessory socket to work, you can either use the battery clamps to connect the inverter or immobilize your steering wheel with a club-type lock. Fully recharging a cell phone or a PDA can take a couple of hours.

Fuel

If there's merely a gas shortage, carpooling can stretch available resources. Sharing rides also tends to establish an informal social network that can work as a support system. The time to establish carpooling is well in advance of any crisis.

Posting a simple chart at each location can help employees self-organize carpools:

- Location
- Home postal code or ZIP code
- Major roads taken to work (in sequence, as if coming to work)
- Name
- Phone
- Ride—Drive—Both
- Starting time at work
- Checkbox for "car full and/or have ride"

Prepare this in a Microsoft Excel spreadsheet, then sort by postal/ZIP code. If you make one day per month carpooling day, with participants eligible for a small prize, a system will form itself.

Some organizations require their senior managers to keep a supply of gas in approved containers at home. The supply should be rotated—empty the can into the car's gas tank and then refill it at a gas station—every three months. That way, senior managers can still make it in if the gas pumps are down. It's a good idea to use fuel stabilizer in stored gas. Otherwise, if rotating is neglected, the old fuel can present a problem. Household fire insurance and condominium bylaws may have their own requirements regarding safe storage.

Light

If the power is out, especially in winter when days are short, lighting will be an issue. Emergency lighting systems in buildings are designed for short outages, usually just a few hours.

If you've ever experienced a blackout you'll realize that, without a bright moon or the spill from streetlights, you can't see your hand in front of your face.

Proximity lighting is an important concept. You don't need to light an entire room, just where you're looking when you work. This stretches lighting resources.

Conventional flashlights and headlamps consume batteries quickly (often within five hours), and filament bulbs burn out after fifteen to twenty hours. A recent development, LED lights, typically run ten times as long on a set of batteries. And LED bulbs are rated to last ten thousand hours.

Outdoor equipment suppliers carry LED headlamps. These both free your hands and put the light exactly where you're looking. Petzl Tikka XP headlamps have a wide-angle feature designed for proximity lighting. They give a perfectly even flood of light (no hot spots) that's ideal for reading and office work.

Two sources for Petzl XP series headlamps:

- Recreational Equipment Incorporated (REI)
- Mountain Equipment Co-Op (MEC)

(See Appendix B for URLs)

If the power fails after dark, an ultra light LED can serve while you find another light. Both Princeton and Photon make key-chain-style LEDs the size of a nickel. The Princeton is widely available (at MEC or REI, or do an Internet search "Pulsar"). The Photon is used by NASA and intelligence agencies.

(See Appendix B for URLs)

The function of a night-light is to help you navigate a dark room. Bicycle taillights use LED technology and make great night-lights. The wide-angle pattern marks an entire room, while the red color preserves night vision. Most taillights will run one hundred hours (equivalent to twenty five-hour emergency candles) on one set of AAA batteries. They are both cheaper and safer than candles.

Water

People can survive for thirty days without food, but only three days without water. In less time than that, electrolyte levels are affected, impairing mental functioning. A person needs three quarts or three liters of water per day to maintain good health. (First aid manuals carry information on sodium and potassium depletion and how to mix an electrolyte solution. Diluted Gatorade can be helpful in replenishing electrolytes.)

(See Appendix B for URLs)

Determine how many people will be onboard for a crisis, and then stock a three-day supply of drinking water—two gallons or eight liters for each.

It's easiest if you do this in small bottles—16 oz. or 500ml. These are often on sale in flats of twenty-four. The small size makes them easy to issue and handle, and people can keep track of how much water they're drinking.

If the power is down, water filtration plants can't operate and a city will use its water reserves, which will last only a few days. When the supply from reservoirs and water towers is gone, there simply is no water—not for drinking and not for fire fighting. Safe LED lights as an alternative to candles then take on a new meaning.

Food

While food is less necessary than we might imagine (humans can survive thirty days without it), it certainly is a comfort. People function much better with it. Persons who are diabetic and those who don't function well when their blood sugar is low will have an especially hard time without food.

Some organizations keep a three-day supply of nonperishable food that doesn't require cooking on hand. At least one large financial institution keeps a two-week supply.

It's important to do a *written* survey of your staff to determine any food allergies. If a person is allergic to peanuts or sulfites or fish oil, the slightest contact with these foods can be life threatening.

Whatever you do, date every package and rotate your stock on a regular basis. Donate the retiring items to a food bank. Your local food bank can tell you items they often need that could work well for your emergency food cache.

Hygiene

Plan on building services not being available. Assemble a list of personal hygiene supplies for your team. Staff members can keep some personal supplies in their desks. Group supplies should include bathroom tissue, soap, paper towels, alcohol gel, and plastic garbage bags. Do some math and decide whether you want a supply for just a few days or longer.

Braun makes an inexpensive, highly rated shaver that runs for sixty minutes on two AA batteries. Do a Google search for "Braun Pocket Twist," and then look at the customer reviews at amazon.com. This can be recommended for personal purchase.

Pandemic influenza and infection control

A contagious disease crisis is a clear possibility. Public health officials maintain that a pandemic is long overdue. You should be prepared for the isolation procedures that hospitals and clinics use to protect their staff and patients.

- The U.S. Homeland Security Council provides a National Strategy for Pandemic Influenza as a free download (seventeen-page PDF).

- The U.S. Department of Health and Human Services manages a Web site that provides one-stop access to U.S. government avian and pandemic flu information. The Planning & Response Activities link leads to annotated subtopics and is highly informative.

- Health Canada has a Web site that lists emergency and disaster resources, including a pandemic information portal.

- The University of Toronto offers a fifteen-point guide with references—"Stand on Guard for Thee: Ethical considerations in preparedness planning for pandemic influenza"—as a PDF file. Topics, such as restricting liberty in the interest of public health, are bookmarked in the PDF.

  (See Appendix B for URLs)

Ask your local public health department about screening procedures and the criteria for getting someone to medical care. Common screening procedures look something like this:

- All personnel disinfect their hands with alcohol gel at the door.
- No one with a temperature of 101°F or 38°C or higher is allowed in.
- No one with general muscle aches or a cough is allowed in.
- Any person who has *any* symptoms must wear a face mask.

During the SARS crisis, one business shut down an office that held 1,400 employees for one week when a worker there developed symptoms. Also during SARS, another business isolated its offices into four sections, each with a separate ventilation system. No one was allowed to travel between sections. That way if one area went under quarantine, the others could still operate.

A public health crisis is something for which you will want to prepare.

Three supplies are required to implement screening:

- Alcohol gel
- Digital fever thermometers (with a good supply of disposable sheaths)
- Face masks

You'll need alcohol-based hand sanitizer for your entrance areas. Sanitizing requires wetting the skin with a 70 percent alcohol solution for twenty seconds. Large hands will need more than small hands. A 16 oz. or 450 ml bottle contains ninety one-teaspoon (5 ml) or forty-five two-teaspoon (10ml) uses. You can buy alcohol gel in large containers and refill smaller dispensing bottles. Multiply people by days and uses to calculate requirements.

A digital fever thermometer is fast, accurate, and easy to read. A good supply of disposable hygienic sleeves for the probe is very important. Your people should also know how to avoid contamination of this equipment. A digital ear thermometer has some advantages. Whatever you use, have your procedures signed off by a qualified medical professional such as a doctor or a registered nurse.

Face masks may all seem the same, but there are critical differences in performance. Dust masks like those sold in hardware stores are useless against biohazards. Inexpensive medical masks—ones that look like a facial tissue with ties—are designed to work for only ten to fifteen minutes—just long enough for a brief visit to a patient's room.

The most effective mask—the one used during the SARS outbreak—is the N95 manufactured by 3M. Each mask costs several dollars and is designed to be effective for up to eight hours. Any mask must be used *strictly* according to instructions; for example, the seal between the mask and face must have absolutely no gaps, and the mask must be changed periodically. No mask is 100 percent effective.

(See Appendix B for URLs)

Regardless of whether there's a health crisis, investing in proper hand washing yields returns in reduced costs associated with common illnesses.

Some online resources:

- Facts about hand washing

- Center for Disease Control (you can download an excellent three-minute video)

- Automated compliance device (outlines procedure and compliance factors)

- History of hand hygiene

- Technical article with references

- Hand washing promotion posters

- Alcohol gels top soap

  (See Appendix B for URLs)

The proper knowledge, together with simple supplies and motivation, can greatly reduce sickness rates and absenteeism.

Home quarantine

Keep in mind that during a public health crisis, thousands of people may be quarantined at home. This usually happens when one family member is identified to be at risk. The entire family is ordered not to leave the house. The exposed person stays in a separate room and is monitored for fever and other symptoms. All family members use alcohol gel and face masks. No family member is allowed out in public—friends or family drop off food and medicine on the doorstep. This continues until the quarantine period is over. The length of quarantine is specific to the illness but can last for weeks.

Data and system security

Computer systems and data are an important part of both business continuity and crisis communication. Performing regular backups is not sufficient. Being able to restore quickly and reliably from a recent backup is the true test.

Approximately 5 percent of system failures occur during attempted restoration from a backup. Experts say many people don't realize how sensitive high-density media can be to shock. And, the tapes they're given to restore from are sometimes useless. One large financial organization, responsible for thousands of pensions, was on number six of seven backups before its system would restore.

Business continuity from an IT perspective is a highly specialized area. One highly regarded resource on this topic is *Disaster Recovery Planning:*

Strategies for Protecting Critical Information Assets by Jon William Tiogo and Jon Toigo.

City resources

Your city may have an emergency management Web site that lists:

- Risks to which your area is especially vulnerable
- Personal emergency preparedness guidelines
- Key agencies involved in disaster response
- Frequently asked questions

Two good examples:

- New York City Office of Emergency Management
- Toronto Office of Emergency Management

    (See Appendix B for URLs)

Find your city's Web site and search it for a similar page on emergency management. If one is not available, you might want to e-mail your mayor's office and point to the above URLs as examples of a public resource you would like for your city.

Planning for human and social needs

It's a mistake to ignore the needs of your staff as individuals and family members. If you plan for their human and social needs, there's a better chance they'll be able to perform their professional roles.

The Institute for Business and Home Safety publishes "Open for Business," a downloadable disaster planning toolkit for organizations with limited resources.

   (See Appendix B for URLs)

HOME

There is more to your staff than their professional roles. In the event of a crisis or a disaster, they need to feel comfortable that their loved ones are safe. Discussing preparedness at home, making a plan, and packing a seventy-two-hour kit will help. Due diligence at home will help ensure their presence at work.

Telus is one company with an excellent public policy on employee and family preparedness.

(See Appendix B for URLs)

Resources for family plans

- READY America—Make a Plan
- Prepare.org—Basic Disaster Plan
- Emergency Preparedness Week—The Plan
- Hurricane Preparedness—Family Disaster Plan
- American Academy of Pediatrics—Children with Special Needs

(See Appendix B for URLs)

Resources for family kits

- READY America—Get a Kit
- Prepare.org—Build Kit
- Emergency Preparedness Week—The Kits
- Hurricane Preparedness—Disaster Supply Kit

(See Appendix B for URLs)

Childcare

Some members of your staff may have young children in day care or in school.

These staff members should have a plan and a letter on file with the school or day care center regarding what to do in an emergency. This may include the name of friends, relatives, and neighbors who are authorized to pick up a child, and what constitutes acceptable identification of these people. You can ask your staff to submit documentation to show that they've done this. Children should know who might pick them up if Mom or Dad has to stay at work.

If yours is a large organization, you may want to look into providing on-site childcare, at least for children of your senior managers who may be away from home for the full duration of an emergency. A local day care center may be able to help. You should learn which resources you can count on—before they're needed.

Elderly parent care

Other members of your staff may have elderly parents who need in-home support on a daily basis.

These staff members can arrange for a relative, neighbor, or friend to help. This should be done as a precaution and tested to see how things go.

It's important to keep detailed notes on medications, what they look like, doses and times. Include the name and number of the pharmacy and prescribing physician.

There should be a medical personnel contact sheet that lists doctors, clinics, hospitals, and visiting nurses. It's also good to have the name of someone who is easily accessible, knows the parent well, and can offer reassurance or help solve an unanticipated problem.

Special needs

At the World Trade Center on September 11, 2001, two men carried a woman in a wheelchair down sixty-eight flights of stairs. Persons of any age can have special needs. There are excellent online resources to help you review relative issues, develop a corporate policy, and put plans and resources into place:

- The Disability Funders Network makes a point ...

    "The needs of people with disabilities following disaster are great and varied—ranging from medical equipment, supplies and medications to batteries for hearing aids, adaptive communications devices and accessible shelter and transportation."

    ... then offers preparedness resources and guidance for funders.

- The American Academy of Pediatrics offers information on emergency preparedness for children with special needs.

- The U.S. Department of Health and Human Services, Office on Disability, offers extensive resources, including a tip sheet for first responders.

- The United States Access Board cites design requirements and links to many resources specific to persons with special needs, including evacuation, assistive products, and planning.

- The American Association on Health and Disability aggregates best practices for emergency preparedness and persons with special needs. These are listed three ways—by type of disaster, type of disability, and role of responder—in a chart that includes format, description, and links.

(See Appendix B for URLs)

Pet care

Pets are just animals ... to people who don't have them. To a pet owner, an eight-year-old dog is family. Worry about *any* family member is a source of stress and distraction. Many of your staff will have pets at home that need to be fed, watered, and cared for.

A neighbor or a nearby friend is the easiest solution. The person needs to be trusted with keys to the house. Arrangements should be made in advance—3:00 AM is not a good time to introduce the concept of emergency pet care to a next-door neighbor.

A note with the key in an envelope referring to an earlier conversation is quick and easy. Work and cell phone numbers should be included, plus details on where everything is. The note is best prepared in advance.

One Canadian city has even included pet care in its emergency shelter planning. Some people will not leave their pets, and this can complicate evacuation and rescue efforts.

You may want to survey your staff to see who has pets and involve them in looking into emergency pet care.

The Center for Disease Control offers an excellent PDF document (three pages) with annotated resources and active links.

   (See Appendix B for URLs)

The American Society for the Prevention of Cruelty to Animals outlines a six-step process for emergency pet preparedness.

   (See Appendix B for URLs)

Including pets in your emergency planning will mitigate one potential problem when you need your full team at work.

◆ ◆ ◆

Emergency Provisions—Questions for discussion

1. What emergency provisions make sense for your organization?
2. Are any emergency provisions already in place?
3. How would you organize requirements by department?
4. How would you budget for emergency provision requirements?

5. Who would manage your emergency provision system?

6. Do you need to work emergency provisions into your policies and procedures?

PART 5
DEVELOPMENT

In a crisis situation, you need a team behind you—a group collectively committed to results. This book was written as a primer and a reference for teams—to help them think about and discuss crisis and disaster situations, and to enable them to pull together to both manage issues and communicate effectively.

14

BUILDING YOUR TEAM

During a crisis your organization will virtually breathe information. There will be constant updates from frontline people on the crisis scene. You'll be feeding information to others within your organization, to important stakeholder groups, and to the media.

It will be a "wicked" situation. Everyone will want answers. And you will be facing urgent deadlines besides other pressures.

Readiness

Every organization exists within the context of an evolving industry and a changing community. This means that crisis communications readiness is an ongoing process.

A solution that sees this as involving only one department—"That's PR's job" or "It belongs to Business Continuity"—just doesn't work. Neither does the approach of offhandedly sending some manager to a conference. You need an entire senior team that's on the same page.

In full crisis mode, everyone in your organization needs to know where to go and what to do. To develop this capability requires a team effort, beginning at the very top. Your senior managers need to collectively own your crisis response. And their reporting relationships need to know how to support them.

You develop a team effort by raising issues, exploring solutions, discussing roles and resources, and process and principles. Discussion will allow your team to clarify its purpose and approach, to organize its goals, and begin working on priorities.

This book was designed to facilitate that process.

A core competence

It's important to think of crisis communications readiness as more than a project. It is a core business competence that needs ongoing management support. You may be the first person in your organization to read this book. If so, you'll want to involve others on your senior management team.

Energy and commitment

When starting a campfire, you begin with tinder before adding kindling and fuel. Tinder catches from a spark and gives you flame. Kindling takes this flame and adds the energy to ignite heavier fuel.

A major fund-raising campaign follows a similar principle. It begins with person-to-person solicitation of lead donors, whose giving sets a standard for others. A second wave of donors, again solicited person to person, further demonstrates support for the cause. It isn't until there is substantial confirmed support—leaders are involved and their commitment is established—that a campaign is ready to go public.

Carefully building a self-sustaining process is important. There is a dominant coalition in every organization that gives the nod to major decisions. To win solid support, you need to involve two or three people within your senior management group. If they agree it's important for the full team to become involved, then it's time for the three or four of you to approach the others.

Flawless Consulting by Peter Block is a useful navigation aid for anyone who would influence change.

Your crisis communications plan will need to permeate your entire organization, at least on a need-to-know basis. Depending on the size of your organization, several management layers may be involved. If your plan filters down with consensus and involvement from the very top, people will receive the signal that they need to get onboard.

Layering builds capability

While it's important to have challenging goals, being overly ambitious can lead to discouragement. One solution is to sketch out layers of reasonable goals. This combines vision with stages of achievement.

An outline with four layers might look like this:

- Layer 1—Initial priorities that everyone agrees to and that can be accomplished within two months using existing budgets—try to involve two or three departments.

- Layer 2—A second tier of priorities that builds on your first layer and is achievable within six months or during the current fiscal year—see if you can involve additional departments.

- Layer 3—Further development in the form of reasonable proposals for next year's budget.

- Layer 4—Where you want to be three years from now, outlined as considerations for department agendas and budget development.

You can discuss these in sequence, or all at once, or you can group the first two and then the second two layers. Any plan should be considered a working document, to be reviewed and revised as you make progress and the process unfolds.

SharedPlan offers elegant and cost-effective software support for project teams. The SharedPlan white paper on team empowerment—LongLive.pdf—is worth reading.

(See Appendix B for URLs)

Gaining purchase

As your team begins to work on your crisis communications plan, some tasks from the first five chapters can provide small wins and help them gain purchase:

- Chapter 1—Key Players and Their Roles—Team members can review their job descriptions in light of how they can support each other in a crisis situation. Human resources can implement changes to job descriptions.

- Chapter 2—Scenarios—Draft a short list of scenarios where you will want resilience. Include examples from several categories: a weather event, a fire or a hazardous materials spill, a power outage, a pandemic situation, criminal misconduct of an employee, a serious technology failure. Assessing your readiness for each will identify issues that are common to several situations.

- Chapter 3—Resources—Working through the Crisis Readiness Checklist is a practical exercise and gives quick results. Two places to begin: index your information resources and produce a directory of your senior team that shows each member's background and job mandate.

- Chapter 4—Roles—Pick one scenario and walk through it, discussing which organizations would be involved, their mandates, and the makeup of your incident management team.

- Chapter 5—Process—Review the Crisis Communications Action Plan and see how it fits with what you've discussed so far. Draft your own action plan that outlines tasks and timelines for key people.

It's important to get off to a solid start, establish ownership with your entire team, and build slowly on achievable goals.

Sub-teams

As your team's work develops, members will need to recruit skills and capacities. They will involve other departments and locations, delegating tasks to sub-teams. Teams work best when they're lean and purpose driven. Give sub-teams room to test their mettle and see how their contribution fits.

This is a major reason that it's so important to get off to a solid start. Crisis management is an ongoing process that incorporates itself into the way your organization operates. Commitment and confidence are essential.

The larger stage

Any organization is part of a community and survives within a network of interdependencies. Those on whom you rely affect your ability to function, and your ability to function affects those who rely on you. The time to meet and manage expectations is *before* a crisis gives assumptions a reality test.

When thinking of teams, consider the dependencies that exist between organizations and the resilience of all the players. Once you have a process started within your own organization, it's time to engage outside organizations and find your place on the larger stage.

Benefits of readiness

Readiness reaps benefits. You will steadily build capabilities and resources. Your team will learn to pull together and members will learn to think on their feet. You will develop new alliances. The process will renew itself and respond to emerging needs.

Even if you never encounter a crisis or a disaster, your organization will benefit because you are committing to excellence in communication.

When a crisis puts you in the spotlight, your performance will include the performance of your team.

◆ ◆ ◆

Building Your Team—Questions for discussion:

1. What would be the likely interest and priorities for each department in the process of building a team?

2. Who would be the first two people in each department to approach about this process?

3. When would it make sense to involve the entire senior team?

4. How could we handle the process of outlining layers of goals?

5. Which initial tasks would make sense for achieving small wins?

6. Who would be the lead manager to coordinate team development?

◆ ◆ ◆

ENDNOTE

An excellent, down-to-earth book on influencing change in organizations is *Flawless Consulting: A Guide to Getting Your Expertise Used* by Peter Block. San Francisco, CA: Jossey-Bass/Pfeiffer, 2000. Harvard and other major universities use Block's book in their organizational development courses.

Appendix A
SUPPLEMENTARY MATERIAL

Here is additional material that will be useful to have on hand in a crisis situation.

Public relations counsel

Professional public relations associations are a good place to start in seeking advice. Read the codes of ethics and information on accreditation, the APR (accredited in public relations), or ABC (accredited business communicator). This will help you find qualified and ethical public relations counsel.

- Canadian Public Relations Society—CPRS
- Public Relations Society of America—PRSA
- Chartered Institute of Public Relations—CIPR (UK)
- International Association of Business Communicators—IABC
- Global Alliance for Public Relations

  (See Appendix B for URLs)

These professional associations are good sources of information on qualified public relations counsel. Contact the chapter of an association that serves your area.

Books on public relations are shelved under call number 659.2 at your local library. If you can't find what you're looking for, ask a librarian if what you need might be available through interlibrary loan.

Handbook of Public Relations by Robert L. Heath (editor) contains four chapters on crisis communications, including a review of some best practices.

See also the **For further reading** list at the end of this appendix for crisis communications titles.

Media request to speak with families *

Journalist _____ staff/freelance
on assignment for _____ print/radio/TV/Web
phone: _____ cellphone: _____
e-mail: _____
date of request _____ story deadline _____
is doing a story on _____
_____
_____

and would like to speak with people who _____
_____
_____

* Please contact this person directly if you are willing to be interviewed.

☐ *See the back of this page for more details.*

(A PDF file for printing this Media Request to Speak with Families is available at www.topstory.ca)

Story Action Sheet

day: _____ date: _____ time: _____
JOURNALIST: _____ staff/freelance
publication/station: _____ print/radio/TV/Web
phone: _____ ext. _____ // fax: _____
Web site: _____ e-mail: _____
other contact info: _____
TOPIC: _____
_____
approach to story: _____
_____
_____
also contacting: _____
_____
DEADLINE: _____ section/program: _____ run date: _____

Contact & research notes

date:	time:	person:	contact, action or follow-through: (*indicates see attached)
_____	_____	_____	_____
_____	_____	_____	_____
_____	_____	_____	_____
_____	_____	_____	_____
_____	_____	_____	_____
_____	_____	_____	_____
_____	_____	_____	_____
_____	_____	_____	_____
_____	_____	_____	_____
_____	_____	_____	_____
_____	_____	_____	_____
_____	_____	_____	_____
_____	_____	_____	_____
_____	_____	_____	_____
_____	_____	_____	_____
_____	_____	_____	_____
_____	_____	_____	_____
_____	_____	_____	_____
_____	_____	_____	_____
_____	_____	_____	_____

(pg. ____ of ____ pgs.) signed _____

(A PDF file for printing this Story Action Sheet is available at
www.topstory.ca)

Microcassette counter vs. minutes

Counter readings versus minutes into tape for Sony and Olympus microcassette recorders at 1.2 cm/sec

Sony	Olympus	minutes	Sony	Olympus	minutes
9	19	1	215	456	31
18	37	2	221	467	32
26	55	3	226	478	33
35	73	4	231	489	34
43	**90**	**5**	**237**	**500**	**35**
51	107	6	242	511	36
59	124	7	247	522	37
67	141	8	252	533	38
74	157	9	257	544	39
82	**173**	**10**	**262**	**554**	**40**
89	189	11	267	565	41
96	204	12	272	575	42
104	219	13	277	586	43
111	234	14	282	596	44
118	**249**	**15**	**287**	**606**	**45**
124	263	16	291	616	46
131	277	17	296	627	47
138	291	18	301	637	48
144	305	19	306	647	49
151	**319**	**20**	**311**	**657**	**50**
157	332	21	316	667	51
163	345	22	320	677	52
169	358	23	325	687	53
175	371	24	330	697	54
181	**384**	**25**	**334**	**707**	**55**
187	396	26	339	717	56
193	408	27	344	727	57
199	420	28	349	737	58
204	432	29	353	747	59
210	**444**	**30**	**358**	**757**	**60**

(A PDF file for printing this Microcassette counter vs. minutes chart* is available at www.topstory.ca)

To determine whether your microcassette has a counter similar to those of Sony or Olympus, note the reading at five minutes using the slow (1.2 cm/sec) tape speed. If it's 43, use the Sony figures. If it's 90, follow those for Olympus. At slow speed, a sixty-minute microcassette tape will record one hour on *each side.*

* My son Peter deserves credit for calibrating this chart and proving that calculus has a practical use.

Public Relations Policy Draft:

Summary

It is important for [organization] to develop mutual, durable relationships that help develop and support its mission. Our communications function must be managed to this purpose.

Universality

We will seek to establish and maintain communication with all groups affected by our policies and actions.

Dialogue

We will improve the quality of relationships with all groups affected by our policies and actions by moving them increasingly toward dialogue.

Scope

The scope of our communications strategy will include all employees and will incorporate excellence in communication management as a vital component of our culture.

Mandate

Responsibility and resources adequate for accomplishing excellence in communications will be mandated to [our organization's] chief executive officer, who may consult with expert counsel and may delegate day-to-day responsibility to a designated public relations coordinator.

Programs

[Organization] will establish programs to:

- Identify and address all issues of public interest
- Develop two-way communication with stakeholder groups
- Find mutual solutions

Values

[Organization's] communications strategy will specifically address critical areas such as:

- Accessibility
- Advocacy
- Anti-racism
- Collaboration
- Customer service
- Efficiency
- Equity
- Public education
- Sustainability

Leadership

One person at the senior staff level will be designated as key contact regarding public relations management.

Media Coordination Policy Draft:

Coordination

A designated senior person will monitor and coordinate and help animate the public relations programs of [organization].

Leading team

The communications coordinator will bring together a team of senior people to set the leading example for other teams—to manage our interdependence with the community through actively sought dialogue and partnership.

Programs

> Central to the communications coordinator's mandate is that communication programs enact the principles outlined in our policy on public relations.

Media

> Contact with the news media (including requests for comment or interviews and issuing of news releases) will normally be directed through the designated communications coordinator.
>
> This is to ensure the following:
>
> - Response time is fast
> - Story lines are identified
> - Background information is available
> - Appropriate spokespeople are involved
> - Staff and volunteers and locations are prepared
> - Consents for publicity are determined
> - Communications advice is available if useful or needed
>
> Our goal is to foster a positive working relationship with journalists and also present the work of [our organization] in a consistent and credible manner.

Materials

> The designated communications coordinator will vet communication materials for general distribution.
>
> This is to ensure the following:
>
> - Proper tone and approach
> - Appropriate use of medium
> - Clearly stated objectives

- Acceptable standards
- Effectiveness

Media directories

Most of the following resources offer Web- or software-based media lists, custom lists, and printed versions of their media lists. The nonprint versions usually involve a subscription period or expiration date, after which the information is no longer accessible.

USA

- Gebbies Press

Canada

- CCNMatthews
- Bowdens MediaSource
- Sources—Media names and numbers

United Kingdom

- Media UK

International

- Mondo-Times—Worldwide media directory

  (See Appendix B for URLs)

A note for small nonprofit organizations with limited budgets:

> Online and software-based directories are no longer accessible once your subscription expires. A print version might be initially inconvenient, but it doesn't expire and can be manually updated. You can build a simple Microsoft Excel database (city, media type, media

name, Web site, newsroom e-mail), then use an inexpensive e-mail verifier program to periodically check which addresses need updating. (These programs verify e-mail addresses without sending a message.) A volunteer can then check the media organization's Web site or place a phone call for the new address. You can find e-mail verifier programs for both Mac and Windows at VersionTracker.com—search "verifier" for your operating system.

(See Appendix B for URLs)

PDAs—Palm

Palm devices and software

If your notebook computer is unavailable, you can access all your contact information and send e-mail with a Palm handheld. Thousands of applications are available.

One Palm device, the Treo, also functions as a cell phone and has a built-in modem.

(See Appendix B for URLs)

FitalyStamp (onscreen keyboard for Palm, PocketPC, and TabletPC devices)

> Software that includes a durable keyboard overlay for the graffiti area. Versions are available for Palm, PocketPC, and TabletPC devices. FitalyStamp allows you to tap forty to fifty words per minute with excellent accuracy.

BackupBuddy VFS

> Palm software that enables you to back up all the files on your Palm to its expansion card and restore all data from the card.

Mark/Space Mail

> A robust e-mail client for the Palm OS that can handle SSL (secure socket layer) and attachments.

Pegasus III (infrared) and BtModem (Bluetooth) modems

> Pegasus III is an infrared fax/data modem that works with serial devices or the infrared port on a Palm device. BtModem is a Bluetooth fax/data modem that works with Bluetooth PDAs. Either will fit in a shirt pocket and can run on AAA batteries.

AllTime

> An elegant and intuitive time tracking application for Palm OS. It generates reports and exports to Microsoft Excel—useful for timing activities during a crisis response test.

TealLock

> A security application for Palm OS. Its features include serial and infrared lockout, 128-bit data encryption for memory and external cards, and a data self-destruct mode to deter brute force attacks. A corporate edition allows administrator controls.

Documents-to-Go

> This application enables Palm users to access Microsoft Word and Excel files.

Dana by AlphaSmart

> The Dana is a Palm laptop with a full-size keyboard that runs for twenty-five hours on three AA batteries. It's designed to endure school use, and its screen is as wide as the text area on letter-size paper. The Dana ships with a suite of productivity applications, and a WiFi version is available.

(See Appendix B for URLs)

PDAs—Blackberry

Many senior government managers are issued Blackberries.

eOffice

> This application enables Blackberry users to access Microsoft Word and Excel files.

PocketMac for Blackberry (freeware)

> This enables a two-way synch between Macs running OSX and Blackberry devices.

> (See Appendix B for URLs)

PDAs—Emergency chargers

BoxWave

> BoxWave makes emergency chargers for handhelds. The Battery Adapter uses AA batteries to recharge a PDA. The VersaChargerPRO is compact and uses either an electrical outlet or vehicle power. (If you are using alkaline batteries with a Dana, you can use one of these to supply power and maintain memory while you change the AAs.)

> (See Appendix B for URLs)

◆ ◆ ◆

Disaster-related associations and resources

Disaster Resource Guide

An excellent and comprehensive source for crisis/emergency management and business continuity information—contains information, vendors, organizations, resources, industry overview, basics, and latest trends.

> (See Appendix B for URLs)

General emergency preparedness sites

- READY.gov—business, America, kids
- American Red Cross—preparedness information on the Web
- Federal Emergency Management Agency—FEMA—United States
- Emergency Preparedness Week—Canada
- American Red Cross—disaster services

  (See Appendix B for URLs)

Specialized resources with Web sites:

- AHDN (All Hands Network)
  Emergency management and business continuity community

- Association of Contingency Planners (ACP)
  A nonprofit trade association dedicated to fostering continued professional growth and development in effective contingency and business resumption planning.

- Canadian Centre for Emergency Preparedness (CCEP)
  A not-for-profit organization devoted to the promotion of emergency risk management to individuals, communities, and organizations—in government and the private sector. Excellent resources on personal and small business preparedness with links to other relevant Web sites.

- Canadian Emergency Preparedness Association (CEPA)
  A national forum to promote emergency preparedness across Canada and represent the interests, aims and opinions of those involved in prevention, planning, response, recovery, and mitigation.

- *Continuity Insights* magazine
  Magazine addresses the need for continuity planning at the highest levels of the organization. Editorial features trends, best practices,

profiles, case studies, timely reporting, product reviews, industry research and data, regulatory affairs, and continuity strategies. An archive lists content going back to 2003 and offers articles in PDF format.

(See Appendix B for URLs)

- *FEMA Emergency Management Guide for Business & Industry*
  A step-by-step approach to emergency planning, response and recovery for companies of all sizes. Sponsored by a public-private partnership with the Federal Emergency Management Agency.

  > "To begin, you need not have in-depth knowledge of emergency management. What you need is the authority to create a plan and a commitment from the chief executive officer to make emergency management part of your corporate culture."

- *Contingency Planning Magazine* (CPM)
  Archives enable you to view listings from past issues of *Contingency Planning & Management* and access full-text magazine content (where available). "The Best of CPM" offers thirteen topic-specific anthologies amassed from past and present issues.

- Disaster Recovery Institute International (DRII)
  Promotes a base of common knowledge for the business continuity planning/disaster recovery industry through education, assistance, and publication of a standard resource base. DRII, with BCI, published the *Professional Practices for Business Continuity Planners* as the industry's international standard. Nearly three thousand business continuity professionals hold DRI International certifications.

- Disaster Recovery Information Exchange (DRIE)
  An organization dedicated to the exchange of ideas within the business continuity and disaster recovery industry—Canadian and Australian chapters.

(See Appendix B for URLs)

- Public Entity Risk Institute (PERI)
  Serves the risk management needs of local governments, small businesses, and small nonprofit entities—clearing house with hundreds of resources, online library, and tools.

- Nonprofit Computer Assisted Risk Evaluation System
  Offers nonprofits a Web-based risk assessment tool with nine modules: introduction to risk management, employment practices, contracts, special events, harm to clients, transportation, internal controls, technology, facilities. Test drive is free. A onetime licensing fee enables you to print a report with specific recommendations.

- International Association of Emergency Managers
  Has a resource page with two dozen selected and annotated URLs, including disaster resources, a best practices network, and colleges and universities offering courses in emergency management.

- American Library Association: Internet resources—Crisis, disaster, and emergency management
  Web sites for researchers—classified and annotated URLs covering:

  - Starting points
  - Emergency management
  - Homeland security
  - Health and medical
  - Terrorism
  - Business continuity and recovery plans
  - Risk assessment and management

- *Edwards Disaster Recovery Directory*
  A commercial directory of business continuity and disaster recovery resources—thousands of listings organized in more than four hundred categories.

  (See Appendix B for URLs)

For further reading

These books contain case studies on crisis communications. All are available through amazon.com and have received excellent reader reviews:

- Barton, Laurence. *Crisis in Organizations: Managing and Communicating in the Heat of Chaos.* Cincinnati, Ohio: South-Western Publishing Co., 1993.

- Coombs, Timothy. *Ongoing Crisis Communication: Planning, Managing and Responding.* Thousand Oaks, California: SAGE Publications, 1999.

- Fearn-Banks, Kathleen. *Crisis Communications: A Casebook Approach.* Mahwah, New Jersey: Lawrence Earlbaum Associates, Inc., 2002/1996.

- Fink, Stephen. *Crisis Management: Planning for the Inevitable.* Lincoln, Nebraska: iUniverse.Inc, 2002/1986.

- Lukaszewski, James A., ABC, APR. *Crisis Communication Plan Components and Models: Crisis Communication Management Readiness.* New York, New York: The Public Relations Society of America, 2005.

References

- Boin A.; Lagadec P.; Michel-Kerjan E.; Overdijk, W. "Critical Infrastructures under Threat: Learning from the Anthrax Scare." *Journal of Contingencies and Crisis Management*, vol 1, no 3, September 2003: 99-104(6). Blackwell Publishing, 2003.

- Katzenbach, Jon R. and Smith, Douglas K. *The Wisdom of Teams: Creating the high performance organization.* New York, New York: HarperBusiness, 2003.

- McKee, Robert. *Story: Substance, Structure, Style, and the Principles of Screenwriting.* New York, New York: Regan Books, Harper Collins, 1997.

- Toigo, Jon William, and Toigo, Jon. *Disaster Recovery Planning: Strategies for Protecting Critical Information Assets.* Upper Saddle River,

New Jersey: Prentice Hall PTR, 2002. Toigo's book for IT professionals, now in its third edition, has a companion Web site.

APPENDIX B
LINKS TO WEB-BASED RESOURCES

This appendix is available as a PDF file for download. *The username and password are on the last page of this book.* The PDF contains bookmarks for chapters and subtopics, and all the URLs are live—you simply click on a URL to go to a site.

I will post an updated PDF file at my Web site www.topstory.ca/crisisteambook.html each year during the first week of January. I can do this only for a limited time. So if you're reading this book five years from now, here are some tips.

URLs have a half-life. There are a number of reasons why a link may not work:

- A busy server may be down for maintenance, or your connection may have timed out.
- The base URL may have changed, or the site may have reorganized its folders.
- The site or the page is no longer available.

It can be worthwhile to try again a few hours later or the next day. If that doesn't work, try a Boolean search that includes the base URL (the part between http://and the next/) in quotes AND a keyword or phrase. You'll want to include words that are certain to be on the Web page, while limiting results.

—*Al Czarnecki, APR*

Part 1: The Team
Chapter 1: Key Players and Their Roles

Problem-based learning

- Samford University has a Web site on problem-based learning that includes links to resources.
  http://www.samford.edu/ctls/problem_based_learning.html

The senior management team

- "Critical Infrastructures under Threat: Learning from the Anthrax Scare"
  http://grace.wharton.upenn.edu/risk/downloads/03-26.pdf

Part 2: The Crisis Soup
Chapter 2: Scenarios

Disasters—broadcast archives

- CBC Archives—Disasters
  http://archives.cbc.ca/IDT-1-70/disasters_tragedies/

- CBS News—Disaster Links
  http://www.cbsnews.com/digitaldan/disaster/disasters.shtml

- BBC—Natural Disasters
  http://www.bbc.co.uk/science/hottopics/naturaldisasters

Disaster information

- FEMA Disaster Hazard Information
  http://www.fema.gov/hazard/

- CDC—Emergency Preparedness and Response
  http://www.bt.cdc.gov/

- Worst United States Disasters
  http://www.infoplease.com/ipa/A0001459.html

- Canadian Disaster Database—information on over seven hundred events
  http://www.ps-sp.gc.ca/res/em/cdd/search-en.asp

- Natural Hazards of Canada—150 year perspective
  http://www.ps-sp.gc.ca/res/em/nh/index-en.asp

- Societal aspects of weather
  http://sciencepolicy.colorado.edu/socasp/toc_img.html

- Engineering disasters—selected Web sites
  http://library.queensu.ca/webeng/guides/disaster/disasters_web.htm

Organizational crises

- Online News Hour—Corporate Ethics
  http://www.pbs.org/newshour/bb/business/ethics/

- Web-miner—Business Ethics
  http://www.web-miner.com/busethics.htm

- The Center for Public Integrity—
  Investigative Journalism in the Public Interest
  http://www.publicintegrity.org/default.aspx
  http://www.publicintegrity.org/icij/

- Endgame Online Research Links
  http://www.endgame.org/links.html

Recommended reading

- "Critical Infrastructures under Threat: Learning from the Anthrax Scare" (4,500 words)
  http://grace.wharton.upenn.edu/risk/downloads/03-26.pdf
- "Lessons Learned or Lessons Forgotten: The Canadian Disaster Experience" (7,400 words)
  http://www.iclr.org/research/publications_policy.htm
  (click on Joe Scanlon, 2001)
- "The Westray Mine Explosion: An Examination of the Interaction Between the Mine Owner and the Media" (12,000 words)
  http://www.cjc-online.ca/viewarticle.php?id=377&layout=html

Chapter 3: Resources

Crisis readiness checklist

The communications policy of the Government of Canada includes an excellent section (item #11) on crisis and emergency communications:

> http://www.tbs-sct.gc.ca/pubs_pol/sipubs/comm/comm02_e.asp

Information useful to record

A one-page form (Story Action Sheet) to capture information for all media queries:

> A PDF file for printing is available at:
> http://www.topstory.ca/crisisteambook.html

Emergency e-mail access

Global Roaming:

> http://international.dialer.net/

Content Management System (CMS)

If you're not familiar with CMS, WebYep is a simple system that offers an online demo:

    http://www.obdev.at/products/webyep/demo.html

Content Management System resources:

    http://www.cmsmatrix.org

    http://www.cmswatch.com/CMS

    http://www.opensourcecms.com

    http://www.webyep.com

HTML kit for cross-trained staff

- htmldog.com—excellent UK site with beginner, intermediate, advanced guides:
  http://www.htmldog.com/about/

- Simple, clear HTML examples
  http://www.w3schools.com/html/html_examples.asp

- Mac OSX freeware HTML editor—Taco HTML Edit
  http://tacosw.com/main.php

- Mac OSX freeware FTP client—CyberDuck FTP
  http://cyberduck.ch/

- Windows freeware HTML editor—Araneae
  http://www.ornj.net/araneae/

- Windows freeware FTP client—Coffee Cup Free FTP
  http://www.coffeecup.com/free-ftp/

- If your regular ISP is not available, a global roaming resource can save the day.
  http://international.dialer.net/

RSS feeds

- MarketingSherpa—Practical news and case studies—RSS Feeds
  http://www.marketingsherpa.com/sample.cfm?contentID=2606
- Wikipedia—RSS (file format)
  http://en.wikipedia.org/wiki/RSS_(protocol)
- Introduction to RSS
  http://www.webreference.com/authoring/languages/xml/rss/intro
- How to feed RSS—A hands-on guide
  http://corz.org/serv/techniques/how-to-feed-rss.php

Portability—USB flash drives

Leading manufacturers of USB flash drives include Kanguru, Kingston Technology, and SanDisk:

http://www.kanguru.com/kanguruusbflash.html (AES model has U.S. FIPS 140-2 Certification)

http://www.kingston.com/flash/datatraveler_home.asp

http://www.sandisk.com/Products/Catalog(1004)-USB_Flash_Drives.aspx

Languages and culture

Foreign Affairs Canada: Centre for Intercultural Learning:

http://www.intercultures.ca/cil-cai/country_insights-en.asp?lvl=8

http://www.dfait-maeci.gc.ca/cfsi-icse/cil-cai/intercultural_glossary-en.asp?lvl=3

http://www.dfait-maeci.gc.ca/cfsi-icse/cil-cai/intercultural_effectiveness-en.asp?lvl=3

Internet Public Library: Ethnicity, Culture, and Race:

http://www.ipl.org/div/subject/browse/soc40.00.00

Chapter 4: Roles

IMS model

Useful URLs on the incident management system (IMS):

    http://www.fema.gov/emergency/nims/index.shtm

    http://training.fema.gov/EMIWeb/IS/is700.asp

    http://training.fema.gov/EMIWeb/downloads/NIMS-Self-Study%20Guide.pdf

    http://www.ec.gc.ca/ee-ue/default.asp

    http://www.env.gov.bc.ca/eemp/

Software relating to incident management system (IMS):

    http://www.incidentcommander.net

    http://www.davislogic.com/command.htm#Command%20Center%20Software

    http://www.ojp.usdoj.gov/nij/pubs-sum/197065.htm

    http://www.incident.com/eoc_design.html

Chapter 5: Process

No URLs in Chapter 5.

Chapter 6: Principles

No URLs in Chapter 6.

Part 3: Issues and Action
Chapter 7: Issues Management

Periodical index

Two periodical indices offer a fertile way to learn how an issue was covered in mainstream magazines:

- *Readers' Guide to Periodicals*
  http://www.hwwilson.com/Databases/Readersg.htm
  http://www.hwwilson.com/Documentation/HowToUseReadersGuide.pdf

- *Canadian Periodical Index*
  http://www.gale.com/customer_service/sample_searches/cpiq.htm

Internet search

Going beyond Google ... explore these two powerful options:

- Clusty (formerly Vivisimo) uses major search engines and clusters results (note the advanced and preferences menus)
  http://clusty.com

- Dogpile searches Google, Yahoo, Ask, MSN (note the advanced search option)
  http://dogpile.com

Blog search engines provide a window to online buzz:

- Technorati lets you create a watchlist
  http://www.technorati.com/

- Blog-maniac offers an excellent orientation to blogs and blogging
  http://www.blog-maniac.com/blogging-articles.htm

Dedicated search software

- Copernic for Windows
  http://www.copernic.com/

- DEVONagent for Mac OSX
  http://www.devon-technologies.com/

Libraries and librarians

IntelliSearch at the Toronto Reference Library:

> http://www.torontopubliclibrary.ca/ask_fee_index.jsp

Focus groups, polls, surveys

- What is an omnibus survey?
  http://www.mrweb.com/whatis.htm

- ORC Macro
  http://www.orcmacro.com/Survey/Surveys/CARAVAN.aspx

- Ipsos Reid
  http://www.ipsos.ca/pa/omni/

- GMI Poll
  http://www.gmi-mr.com/gmipoll/

- Survey software directory
  http://www.capterra.com/survey-software

- "Conducting Web-based Surveys" (article)
  http://pareonline.net/getvn.asp?v=7&n=19

- SurveyMonkey (one example of a Web-based service)
  http://surveymonkey.com/.

Business ethics—online resources

- Ethics Resource Center
  http://www.ethics.org/

- Center for Ethical Business Cultures
  http://www.cebcglobal.org/
- International Business Ethics Institute (IBEI)
  http://www.business-ethics.org/primer.asp
  http://www.business-ethics.org/links.asp
  http://www.business-ethics.com/resources/business_ethics_directory.html
- *Business Ethics* magazine
  http://www.business-ethics.com/
- Corporate Governance
  http://www.corpgov.net/
- *Complete Guide to Ethics Management*
  http://www.managementhelp.org/ethics/ethxgde.htm

Chapter 8: Emotional Intelligence

Approach

Once that's been done, a useful precept of TA (transactional analysis) is that an "I'm OK, you're OK," approach works. People generally respond to respect by returning it, and a rational approach tends to engender the same.

> http://www.businessballs.com/transact.htm
> http://www.itaa-net.org
> http://www.iee.org/oncomms/sector/management/transactional_analysis.cfm

Chapter 9: Focus

No URLs in Chapter 9

Chapter 10: News Conferences

Microcassette recorders

If you're using rechargeable batteries, the MAHA brand is consistently recognized as a leader.

http://www.mahaenergy.com/store/Index.asp

A new generation of rechargeable batteries greatly reduces self-discharge.

http://www.eneloopusa.com/home.html

http://eneloop.ca/

http://www.panasonic.ca/english/batteries/batteries_chargers/chargers/index.asp#

Recording phone interviews with the consent of at least one party to the call is usually legal in the United States and Canada.

http://www.callcorder.com/phone-recording-law-america.htm

http://www.callcorder.com/phone-recording-law-canada.htm

MP3 files

As of this writing, the Edirol R-09 and the Marantz PMD660 are two professional MP3 recorders that meet these criteria.

http://www.amazon.com/gp/product/customer-reviews/B000FPQFKO/

http://www.amazon.com/gp/product/customer-reviews/B00091R29C/

http://www.transom.org/tools/recording_interviewing/

Intuitive and inexpensive shareware—MP3 Trimmer for Mac OSX and MP3 EasySplitter for Windows—allows you to review MP3 files and extract sound bytes without losing quality.

http://deepniner.net/mp3trimmer/

http://www.versiontracker.com/dyn/moreinfo/macosx/20938

http://www.softdd.com/mp3-splitter/index.html

http://www.versiontracker.com/dyn/moreinfo/win/29317

Chapter 11: Media Accuracy

Monitoring the media

Many suppliers of media lists offer a monitoring service. You'll also find resources on the Web.

http://dmoz.org/News/Media/Services/Media_Monitoring

If you don't subscribe to a monitoring service, these resources can help you track coverage.

Online news searches:

- Journalism.net—Canada and United States pages
  http://journalismnet.com/

- American Journalism Review—tracks multiple sources
  http://www.ajr.org/Newspapers.asp?MediaType=1

- NewsLink—U.S. and Canadian media
  http://newslink.org/

- Google News enables you to search and browse 4,500 news sources
  http://news.google.com/

- Yahoo News offers a keyword search across multiple media channels
  http://news.yahoo.com/

- Disaster Center—newspaper, television, radio links—United States
  http://www.disastercenter.com/

Broadcast URLs—streaming media feeds:

- Radio and television stations on the Web (worldwide)
  http://radiostationworld.com/stations_on_the_web/default.asp

- Radio Locator (United States and Canada)
  http://radio-locator.com/cgi-bin/page?page=about

Print URLs:

- Newspaper Association of America—rich resources
  http://www.naa.org/

- Canadian Newspaper Association—rich and varied resources
  http://www.cna-acj.ca/

- World Press Review—perspectives from around the globe
  http://worldpress.org/

Google Alerts

Google Alerts can work as a simple, worldwide media monitoring service.

http://www.google.com/alerts

http://www.google.com/alerts/faq.html?hl=en

Windows users can identify recurring phrases in media coverage by using Textanz.

http://www.cro-code.com/textanz.jsp

http://www.versiontracker.com/dyn/moreinfo/win/62835

If you have an Intel-based Mac, you can run Windows applications without rebooting.

http://www.apple.com/getamac/windows.html

http://www.parallels.com/en/products/desktop/

DNA13

DNA13 offers very powerful (and priced accordingly) real-time issues and media monitoring software that's used by some of the world's largest corporations.

    http://dna13.com/products.html

    http://dna13.com/customers.html

    http://dna13.com/partners.html

Facilitating media accuracy

The BBC makes some excellent points on how to develop a relationship with the media:

- http://www.bbc.co.uk/connectinginacrisis/12.shtml

Stylebooks and codes of ethics will give you a sense for media standards:

- *Associated Press Stylebook and Broadcast News Handbook*
  http://www.apstylebook.com/

- *The Canadian Press Stylebook and Broadcast News Talk*
  http://www.cp.org/books.aspx?id=182
  http://www.apbookstore.com/apbroadnewha.html
  products > books > CP Style Book
  products > books > BN News Talk

- Media Codes of Ethics—Canada, United States, and other countries
  http://www.ijnet.org/Director.aspx?P=Ethics

- Ethics in Public Broadcasting (United States)
  http://www.current.org/ethics/

Journalists and editors can make honest mistakes. Assume they act in good faith:

- Regret the error
  http://www.regrettheerror.com/

- *American Journalism Review*: Errors on air
  http://www.ajr.org/article.asp?id=3094

- *Columbia Journalism Review*: Handling corrections
  http://archives.cjr.org/year/99/4/poll.asp

- *American Society of Newspaper Editors*: Credibility
  http://www.asne.org/kiosk/editor/01.march/bressers1.htm

Chapter 12: Testing

Time tracking software

TimeBox is a simple, inexpensive, Java-based time tracking application:

http://taubler.com/timebox/newuser.shtml#dl

For larger organizations, it integrates with ClokBox and TimeSuite:

http://taubler.com/clokbox
http://taubler.com/timesuite

For Palm users, AllTime is an excellent and intuitive solution that exports to Excel:

http://www.iambic.com/alltime/palmos

Part 4: Resilience and Continuity
Chapter 13: Emergency Provisions

OFFICE

Employee Assistance

Surviving and Thriving: Living Through a Traumatic Experience, an audio book, is available on CD or for download on the Web:

http://www.aaets.org/survivedownload.htm

These URLs offer a good overview of EAP programs:

> http://www.opm.gov/Employment_and_Benefits/WorkLife/Health Wellness/EAP/index.asp
>
> http://www.nimh.nih.gov
>
> http://www.eapassn.org/public/pages/index.cfm?pageid=1
>
> http://www.familyserviceseap.com/home/index_e.html

Eaposters.com gives you a sense of how an EAP program can be promoted through posters in the workplace:

> http://www.eaposters.com

Specialized resources relating to trauma and disaster mental health:

- National Association of School Psychologists
  http://www.nasponline.org/resources/crisis_safety/
- CDC—Disaster mental health
  http://www.bt.cdc.gov/mentalhealth/
- National Mental Health Information Center—
  Disaster mental health recovery services
  http://www.mentalhealth.samhsa.gov/cmhs/EmergencyServices/
- American Academy of Experts in Traumatic Stress
  http://www.aaets.org/
- Red Cross > Masters of Disaster > Facing Fear (lesson plan for schools)
  http://www.redcross.org/disaster/masters/facingfear/

Commuting issues

Telework/telecommuting:

> http://www.telcoa.org
> http://www.ivc.ca
> http://www.telework.gov
> http://www.flexibility.co.uk

Internet-based fax services:

>http://www.efax.com
>http://www.myfax.com
>http://www.trustfax.com
>http://www.instantinfo.com/jump_pages/rightfax.html

Self-inflating foam mats:

>http://www.rei.com
>http://www.mec.ca

>(search "self inflating")

Power

APC fuel cell

>http://www.apcc.com/products/family/index.cfm?id=285

Xantrex inverters

>http://www.xantrex.com/web/id/179/p/1/pt/32/product.asp
>http://www.xantrex.com/web/id/186/p/1/pt/32/product.asp

Light

Two sources for Petzl XP series headlamps (search "LED XP"):

- Recreational Equipment Incorporated (REI)
  http://www.rei.com

- Mountain Equipment Co-Op (MEC)
  http://www.mec.ca

The Princeton is widely available (see above URLs—at MEC or REI, search "Pulsar").

The Photon is used by NASA and intelligence agencies.

>http://www.photonlight.com/products/photon_microlight.html

Water

People can survive thirty days without food, but only three days without water.

> http://www.feinberg.northwestern.edu/nutrition/factsheets/water.html

Pandemic influenza and infection control

- The U.S. Homeland Security Council provides a National Strategy for Pandemic Influenza as a free download (17-page PDF).
  http://www.ntis.gov/products/strat-pandemic.asp

- The Department of Health and Human Services manages a Web site with one-stop access to U.S. government avian and pandemic flu information. The Planning & Response Activities link leads to annotated subtopics and is highly informative.
  http://www.pandemicflu.gov/
  http://www.pandemicflu.gov/plan/

- Health Canada has a Web site with emergency and disaster resources, including a pandemic information portal.
  http://www.hc-sc.gc.ca/ed-ud/index_e.html
  http://www.influenza.gc.ca/

- The University of Toronto offers a fifteen-point guide with references—"Stand on Guard for Thee: Ethical considerations in preparedness planning for pandemic influenza"—as a PDF file. Topics, such as restricting liberty in the interest of public health, are bookmarked in the PDF.
  http://www.utoronto.ca/jcb/home/documents/pandemic.pdf

3M N95 mask:

> http://www.3m.com/
> Click on your country—search "health care N95"—click on Health care N95 link You can download PDF files regarding applications and proper use.

Facts about hand washing

> http://healthlink.mcw.edu/article/955074416.html

Center for Disease Control (you can download an excellent three-minute video)

> http://www.cdc.gov/ncidod/op/handwashing.htm

Automated compliance device (outlines procedure and compliance factors)

> http://www.hygenius.com/faqs.htm

History of hand hygiene

> http://www.accessexcellence.org/AE/AEC/CC/hand_background.html

Technical article with references

> http://www.hi-tm.com/Documents/Safehands.html

Hand washing promotion posters

> http://www.visualhealthresources.com/index.htm

Alcohol gels top soap

> http://preventdisease.com/news/articles/alcohol_gels_tops_hospitals.shtml

City resources

- New York City Office of Emergency Management
  http://www.nyc.gov/html/oem/html/home/home.shtml
  http://www.nyc.gov/html/oem/html/ready/ready.shtml

- Toronto Office of Emergency Management
  http://www.toronto.ca/wes/techservices/oem/index.htm
  http://www.toronto.ca/wes/techservices/oem/risks.htm
  http://www.toronto.ca/wes/techservices/oem/links.htm

Planning for human and social needs

The Institute for Business and Home Safety publishes *Open for Business,* a downloadable disaster planning toolkit for organizations with limited budgets.

> http://www.ibhs.org/docs/openforbusiness.pdf

HOME

Family preparedness

Telus is one example of an excellent public policy on employee and family preparedness:

> http://about.telus.com/publicpolicy/prep_employees.html

Resources for family plans

> http://www.ready.gov/america/makeaplan/index.html
> http://www.prepare.org/basic/DisasterPlan.pdf
> http://www.epweek.ca/index_e.asp
> http://www.nhc.noaa.gov/HAW2/english/prepare/family_plan.shtml
> http://www.aap.org/healthtopics/specialneeds.cfm

Resources for family kits

> http://www.ready.gov/america/getakit/index.html
> http://www.prepare.org/basic/basic.htm
> http://www.getprepared.ca/kit/kit_e.asp
> http://www.nhc.noaa.gov/HAW2/english/prepare/supply_kit.shtml

Special needs

The Disability Funders Network

> http://www.disabilityfunders.org/epdr.html

The American Academy of Pediatrics offers information on emergency preparedness for children with special needs:

http://www.aap.org/healthtopics/specialneeds.cfm

The U.S. Department of Health and Human Services, Office on Disability:

http://www.hhs.gov/od/emergencypreparedness.html
http://www.access-board.gov/evac.htm

The American Association on Health and Disability aggregates best practices:

http://www.aahd.us/page.php?pname=health/research/best_practices/emergency

Pet care

The Center for Disease Control offers a PDF document with annotated resources and active links:

http://www.bt.cdc.gov/disasters/petprotect.asp

The American Society for the Prevention of Cruelty to Animals outlines a six-step process for emergency pet preparedness:

http://www.aspca.org/site/PageServer?pagename=pets_emergency

Part 5: Development
Chapter 14: Building Your Team

Layering builds capability

SharedPlan application support for project teams

http://www.sharedplan.com/index.html
http://www.sharedplan.com/planleader.html
http://www.sharedplan.com/project_server.html

http://www.sharedplan.com/openplanninglite.html
http://www.sharedplan.com/downloads/LongLive.pdf

Appendix A

Public relations counsel

Global Alliance for Public Relations

> http://www.globalpr.org
> http://www.globalpr.org/knowledge/ethics.asp

Canadian Public Relations Society

> http://www.cprs.ca
> http://cprs.ca/Accreditation/e_accreditation.htm
> http://cprs.ca/AboutCPRS/e_code.htm

Public Relations Society of America

> http://www.prsa.org
> http://www.prsa.org/PD/apr/index.html
> http://www.prsa.org/aboutUs/ethics/index.html

Chartered Institute of Public Relations (UK)

> http://www.cipr.co.uk
> http://www.cipr.co.uk/qualifications/index.htm
> http://www.cipr.co.uk/direct/membership.asp?v1=code

International Association of Business Communicators

> http://www.iabc.com
>
> http://www.iabc.com/abc
>
> http://www.iabc.com/about/code.htm

Media directories

USA

- Gebbies Press
  http://www.gebbieinc.com/

Canada

- CCN Matthews
  http://www.cdn-news.com/news/releases/medialist.jsp
- Bowdens MediaSource
  http://www.bowdens.com/SOLUTIONS/medianamerican.htm
- Sources—Media names and numbers
  http://www.sources.com/mnn/

United Kingdom

- Media UK
  http://www.mediauk.com/

International

- Mondo-Times—Worldwide media directory
  http://www.mondotimes.com/

A note for small nonprofit organizations with limited budgets:

E-mail verifier programs for both Mac and Windows at VersionTracker
   http://versiontracker.com

Search "verifier" for your operating system.

PDAs—Palm

Palm devices and software
> http://www.palm.com

Palm Treo
> http://store.palm.com/home/index.jsp

FitalyStamp (onscreen keyboard for Palm, PocketPC and TabletPC devices)
> http://www.fitaly.com/product/fitalystamp.htm

BackupBuddy VFS
> http://www.bluenomad.com/bbvfs/prod_bbvfs_details.html

Mark/Space Mail
> http://www.markspace.com/mail.html

Pegasus III (infrared) and BtModem (Bluetooth) modems
> http://www.3jtech.com

AllTime
> http://www.iambic.com/alltime/palmos

TealLock
> http://www.tealpoint.com/softlock.htm

Documents-to-Go
> http://www.dataviz.com

Dana by AlphaSmart
> http://www3.alphasmart.com/products/dana-w_ln.html
> http://www1.alphasmart.com/danastore/pegasus.html

PDAs—Blackberry

http://www.blackberry.com

eOffice

https://dynoplex.pdapointer.com/view/userProducts.php?userID=180

PDAs—Emergency chargers

BoxWave

http://www.boxwave.com/products/batteryadapter
http://www.boxwave.com/products/versachargerpro/index.htm
http://www.boxwave.com/products/easyfinder/index.htm

◆ ◆ ◆

Disaster-related associations and resources

Disaster Resource Guide

http://disaster-resource.com

General emergency preparedness sites

http://www.ready.gov
http://www.prepare.org
http://www.fema.gov/library
http://www.emergencypreparednessweek.ca
http://www.redcross.org/services/disaster

AHDN—Emergency management and business continuity community

http://www.all-hands.net

Association of Contingency Planners (ACP)

http://acp-international.com

Canadian Centre for Emergency Preparedness (CCEP)
   http://www.ccep.ca

Canadian Emergency Preparedness Association (CEPA)
   http://www.cepa-acpc.ca

Continuity Insights Magazine
   http://continuityinsights.com/

FEMA Emergency Management Guide For Business & Industry
   http://www.fema.gov/business/guide/index.shtm

Contingency Planning Magazine (CMP)
   http://www.contingencyplanning.com

Disaster Recovery Institute International (DRII)
   http://drii.org/DRII/

An archive of past shows includes both audio and RSS podcast feeds:
   http://wmet1160.com/artman/publish/businesscontinuitymatters.shtml#audio
   http://wmet1160.com/artman/publish/businesscontinuitymatters.shtml#podcast

Disaster Recovery Information Exchange (DRIE)
   http://drie.org

Public Entity Risk Institute (PERI)
   http://www.riskinstitute.org/peri/

Nonprofit Computer Assisted Risk Evaluation System
   http://nonprofitrisk.org/cares/cares.htm

International Association of Emergency Managers

http://www.iaem.com/resources/links/intro.htm

American Library Association: Internet resources—Crisis, disaster, and emergency management

http://www.ala.org/ala/acrl/acrlpubs/crlnews/backissues2002/novmonth/crisisdisaster.htm

Edwards Disaster Recovery Directory

http://www.EdwardsInformation.com/default1.html

Access to PDF files

The PDF of Appendix B is located at http://topstory.ca/crisisteambook.html. You'll find other support files available for download there as well.

To enter the private area you'll need:

username: team
password: read

These are case sensitive—they must be all lowercase.

Access to this protected area is an exclusive benefit for readers who have purchased this book. Please respect the author's intellectual property and do not give out this access information.

978-0-595-40613-5
0-595-40613-0

Printed in the United States
131720LV00009B/1/A